AF545257

Introduction to Computer Performance Analysis with Mathematica

This is a volume in
COMPUTER SCIENCE AND SCIENTIFIC COMPUTING

Werner Rheinboldt, editor

Introduction to Computer Performance Analysis with Mathematica

Arnold O. Allen
Software Technology Division
Hewlett-Packard
Roseville, California

AP PROFESSIONAL
Harcourt Brace & Company, Publishers

Boston San Diego New York
London Sydney Tokyo Toronto

This book is printed on acid-free paper. 

AP PROFESSIONAL
955 Massachusetts Avenue, Cambridge, MA 02139

An Imprint of ACADEMIC PRESS, INC.
A Division of HARCOURT BRACE & COMPANY

United Kingdom Edition published by
ACADEMIC PRESS LIMITED
24–28 Oval Road, London NW1 7DX

ISBN 0-12-051070-7

Printed in the United States of America
93 94 95 96 EB 9 8 7 6 5 4 3 2 1

For my son, John,
and my colleagues
at the Hewlett-Packard
Software Technology Division

LIMITED WARRANTY AND DISCLAIMER OF LIABILITY

ACADEMIC PRESS PROFESSIONAL (APP) AND ANYONE ELSE WHO HAS BEEN INVOLVED IN THE CREATION OR PRODUCTION OF THE ACCOMPANYING SOFTWARE AND MANUAL (THE "PRODUCT") CANNOT AND DO NOT WARRANT THE PERFORMANCE OR RESULTS THAT MAY BE OBTAINED BY USING THE PRODUCT. THE PRODUCT IS SOLD "AS IS" WITHOUT WARRANTY OF ANY KIND (EXCEPT AS HEREAFTER DESCRIBED), EITHER EXPRESSED OR IMPLIED, INCLUDING, BUT NOT LIMITED TO, ANY WARRANTY OF PERFORMANCE OR ANY IMPLIED WARRANTY OF MERCHANTABILITY OR FITNESS FOR ANY PARTICULAR PURPOSE. APP WARRANTS ONLY THAT THE MAGNETIC DISKETTE(S) ON WHICH THE SOFTWARE PROGRAM IS RECORDED IS FREE FROM DEFECTS IN MATERIAL AND FAULTY WORKMANSHIP UNDER NORMAL USE AND SERVICE FOR A PERIOD OF NINETY (90) DAYS FROM THE DATE THE PRODUCT IS DELIVERED. THE PURCHASER'S SOLE AND EXCLUSIVE REMEDY IN THE EVENT OF A DEFECT IS EXPRESSLY LIMITED TO EITHER REPLACEMENT OF THE DISKETTE(S) OR REFUND OF THE PURCHASE PRICE, AT APP'S SOLE DISCRETION.

IN NO EVENT, WHETHER AS A RESULT OF BREACH OF CONTRACT, WARRANTY OR TORT (INCLUDING NEGLIGENCE), WILL APP BE LIABLE TO PURCHASER FOR ANY DAMAGES, INCLUDING ANY LOST PROFITS, LOST SAVINGS OR OTHER INCIDENTAL OR CONSEQUENTIAL DAMAGES ARISING OUT OF THE USE OR INABILITY TO USE THE PRODUCT OR ANY MODIFICATIONS THEREOF, OR DUE TO THE CONTENTS OF THE SOFTWARE PROGRAM, EVEN IF APP HAS BEEN ADVISED OF THE POSSIBILITY OF SUCH DAMAGES, OR FOR ANY CLAIM BY ANY OTHER PARTY.

SOME STATES DO NOT ALLOW LIMITATION ON HOW LONG AN IMPLIED WARRANTY LASTS, NOR EXCLUSIONS OR LIMITATIONS OF INCIDENTAL OR CONSEQUENTIAL DAMAGES, SO THE ABOVE LIMITATIONS AND EXCLUSIONS MAY NOT APPLY TO YOU. THIS WARRANTY GIVES YOU SPECIFIC LEGAL RIGHTS, AND YOU MAY ALSO HAVE OTHER RIGHTS WHICH VARY FROM JURISDICTION TO JURISDICTION.

THE RE-EXPORT OF UNITED STATES ORIGIN SOFTWARE IS SUBJECT TO THE UNITED STATES LAWS UNDER THE EXPORT ADMINISTRATION ACT OF 1969 AS AMENDED. ANY FURTHER SALE OF THE PRODUCT SHALL BE IN COMPLIANCE WITH THE UNITED STATES DEPARTMENT OF COMMERCE ADMINISTRATION REGULATIONS. COMPLIANCE WITH SUCH REGULATIONS IS YOUR RESPONSIBILITY AND NOT THE RESPONSIBILITY OF APP.

Contents

Preface

When you can measure what you are speaking about and express it in numbers you know something about it; but when you cannot express it in numbers, your knowledge is of a meager and unsatisfactory kind.
Lord Kelvin

In learning the sciences, examples are of more use than precepts.
Sir Isaac Newton

Make things as simple as possible but no simpler.
Albert Einstein

This book has been written as a beginner's guide to computer performance analysis. For those who work in a predominantly IBM environment the typical job titles of those who would benefit from this book are Manager of Performance and Capacity Planning, Performance Specialist, Capacity Planner, or System Programmer. For Hewlett-Packard installations job titles might be Data Center Manager, Operations Manager, System Manager, or Application Programmer. For installations with computers from other vendors the job titles would be similar to those from IBM and Hewlett-Packard.

In keeping with Einstein's principle stated above, I tried to keep all explanations as simple as possible. Some sections may be a little difficult for you to comprehend on the first reading; please reread, if necessary. Sometimes repetition leads to enlightenment. A few sections are not necessarily hard but a little boring as material containing definitions and new concepts can sometimes be. I have tried to keep the boring material to a minimum.

This book is written as an interactive workbook rather than a reference manual. I want you to be able to try out most of the techniques as you work your way through the book. This is particularly true of the performance modeling sections. These sections should be of interest to experienced performance analysts as well as beginners because we provide modeling tools that can be used on real systems. In fact we present some new algorithms and techniques that were developed at the Hewlett-Packard Performance Technology Center so that we could model complex customer computer systems on IBM-compatible Hewlett-Packard Vectra computers.

Anyone who works through all the examples and exercises will gain a basic understanding of computer performance analysis and will be able to put it to use in computer performance management.

The prerequisites for this book are a basic knowledge of computers and some mathematical maturity. By basic knowledge of computers I mean that the reader is familiar with the components of a computer system (CPU, memory, I/O devices, operating system, etc.) and understands the interaction of these components to produce useful work. It is not necessary to be one of the digerati (see the definition in the Definitions and Notation section at the end of this preface) but it would be helpful. For most people mathematical maturity means a semester or so of calculus but others reach that level from studying college algebra.

I chose *Mathematica* as the primary tool for constructing examples and models because it has some ideal properties for this. Stephen Wolfram, the original developer of *Mathematica*, says in the "What is *Mathematica*?" section of his book [Wolfram 1991]:

Mathematica is a general computer software system and language intended for mathematical and other applications.

You can use *Mathematica* as:

1. A numerical and symbolic calculator where you type in questions, and *Mathematica* prints out answers.

2. A visualization system for functions and data.

3. A high-level programming language in which you can create programs, large and small.

4. A modeling and data analysis environment.

5. A system for representing knowledge in scientific and technical fields.

6. A software platform on which you can run packages built for specific applications.

7. A way to create interactive documents that mix text, animated graphics and sound with active formulas.

8. A control language for external programs and processes.

9. An embedded system called from within other programs.

Mathematica is incredibly useful. In this book I will be making use of a number of the capabilities listed by Wolfram. To obtain the maximum benefit from this book I strongly recommend that you work the examples and exercises using the *Mathematica* programs that are discussed and that come with this book. Instructions for installing these programs are given in Appendix A.

Although this book is designed to be used interactively with *Mathematica*, any reader who is interested in the subject matter will benefit from reading this book and studying the examples in detail without doing the *Mathematica* exercises.

You need not be an experienced *Mathematica* user to utilize the programs used in the book. Most readers not already familiar with *Mathematica* can learn all that is necessary from "What is *Mathematica?*" in the Preface to [Wolfram 1991], from which we quoted above, and the "Tour of *Mathematica*" followed by "*Mathematica* Graphics Gallery" in the same book.

For those who want to consider other *Mathematica* books we recommend the excellent book by Blachman [Blachman 1992]; it is a good book for both the beginner and the experienced *Mathematica* user. The book by Gray and Glynn [Gray and Glynn 1991] is another excellent beginners' book with a mathematical orientation. Wagon's book [Wagon 1991] provides still another look at how *Mathematica* can be used to explore mathematical questions. For those who want to become serious *Mathematica* programmers, there is the excellent but advanced book by Maeder [Maeder 1991]; you should read Blachman's book before you tackle this book. We list a number of other *Mathematica* books that may be of interest to the reader at the end of this preface. Still others are listed in Wolfram [Wolfram 1991].

We will discuss a few of the elementary things you can easily do with *Mathematica* in the remainder of this preface.

Mathematica will let you do some recreational mathematics easily (some may consider "recreational mathematics" to be an oxymoron), such as listing the first 10 prime numbers. (Recall that a prime number is an integer that is divisible only by itself and one. By convention, 2 is the smallest positive prime.)

`Table` generates a set of primes. `Prime[i]` generates the `i`th prime number.	`In[5]:= Table[Prime[i], {i, 10}]`
Voilà! The primes.	`Out[5]= {2, 3, 5, 7, 11, 13, 17, 19, 23, 29}`

If you want to know what the millionth prime is, without listing all those preceding it, proceed as follows.

What is the millionth prime? This is it!	In[7]:= Prime[1000000] Out[7]= 15485863

You may be surprised at how small the millionth prime is.
You may want to know the first 30 digits of π. (Recall that π is the ratio of the circumference of a circle to its diameter.)

Pi is the **Mathematica** word for π. This is 30 digits of π!	In[4]:= N[Pi, 30] Out[4]= 3.14159265358979323846264338328

The number π has been computed to over two billion decimal digits. Before the age of computers an otherwise unknown British mathematician, William Shanks, spent twenty years computing π to 707 decimal places. His result was published in 1853. A few years later it was learned that he had written a 5 rather than a 4 in the 528th place so that all the remaining digits were wrong. Now you can calculate 707 digits of π in a few seconds with *Mathematica* and all 707 of them will be correct!

Mathematica can also eliminate much of the drudgery we all experienced in high school when we learned algebra. Suppose you were given the messy expression $6x^2y^2 - 4xy^3 + x^4 - 4x^3y + y^4$ and told to simplify it. Using *Mathematica* you would proceed as follows:

```
In[3]:= 6 x^2 y^2 - 4 x y^3 + x^4 -4 x^3 y + y^4

          4      3         2  2        3    4
Out[3]= x  - 4 x  y + 6 x  y  - 4 x y  + y

In[4]:= Simplify[%]

                 4
Out[4]= (-x + y)
```

If you use calculus in your daily work or if you have to help one of your children with calculus, you can use *Mathematica* to do the tricky parts. You may remember the scene in the movie *Stand and Deliver* where Jaime Escalante of James A. Garfield High School in Los Angeles uses tabular integration by parts to show that

$$\int x^2 \sin x dx = -x^2 \cos x + 2x \sin x + 2\cos x + C$$

With *Mathematica* you get this result as follows.

This is the **Mathematica** command to integrate. **Mathematica** gives the result this way. The floating 2 is the exponent of x.	`In[6]:= Integrate[x^2 Sin[x], x]`
	`                    2`
	`Out[6]= (2 - x  ) Cos[x] + 2 x`
	`Sin[x]`

Mathematica can even help you if you've forgotten the quadratic formula and want to find the roots of the polynomial $x^2 + 6x - 12$. You proceed as follows:

```
In[4]:= Solve[x^2 + 6 x -12==0, x]

                 -6 + 2 Sqrt[21]             -6 - 2 Sqrt[21]
Out[4]= {{x -> ---------------}, {x -> ---------------
}}
                        2                          2
```

None of the above Mathematica output looks exactly like what you will see on the screen but is as close as I could capture it using the SessionLog.m functions.

We will not use the advanced mathematical capabilities of *Mathematica* very often but it is nice to know they are available. We will frequently use two other powerful strengths of *Mathematica*. They are the advanced programming language that is built into *Mathematica* and its graphical capabilities.

In the example below we show how easy it is to use *Mathematica* to generate the points needed for a graph and then to make the graph. If you are a beginner to computer performance analysis you may not understand some of the parameters used. They will be defined and discussed in the book. The purpose of this example is to show how easy it is to create a graph. If you want to reproduce the graph you will need to load in the package **work.m**. The *Mathematica* program **Approx** is used to generate the response times for workers who are using terminals as we allow the number of user terminals to vary from 20 to 70. We assume there are also 25 workers at terminals doing another application on the computer system. The vector *Think* gives the think times for the two job classes and the array *Demands* provides the service requirements for the job classes. (We will define think time and service requirements later.)

Generates the basic service data
Sets the population sizes
Sets the think times
Plots the response times versus the number of terminals in use.
	`demands = {{ 0.40, 0.22}, { 0.25, 0.03 } }`
	`pop = { 50, 25 }`
	`think = { 30, 45 }`
	
	`Plot[ Approx[ { n, 20 }, think, demands, 0.0001 ][[1,1]], { n, 10, 70 } ]`

This is the graph produced by the plot command.	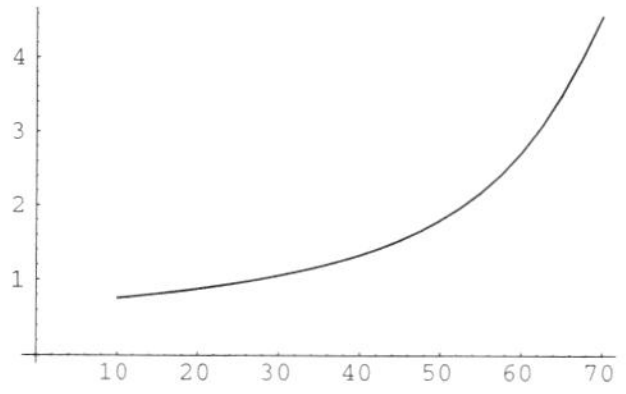

Acknowledgments

Many people helped bring this book into being. It is a pleasure to acknowledge their contributions. Without the help of Gary Hynes, Dan Sternadel, and Tony Engberg from Hewlett-Packard in Roseville, California this book could not have been written. Gary Hynes suggested that such a book should be written and provided an outline of what should be in it. He also contributed to the *Mathematica* programming effort and provided a usable scheme for printing the output of *Mathematica* programs—piles of numbers are difficult to interpret! In addition, he supplied some graphics and got my workstation organized so that it was possible to do useful work with it. Dan Sternadel lifted a big administrative load from my shoulders so that I could spend most of my time writing. He arranged for all the hardware and software tools I needed as well as FrameMaker and *Mathematica* training. He also handled all the other difficult administrative problems that arose. Tony Engberg, the R & D Manager for the Software Technology Division of Hewlett-Packard, supported the book from the beginning. He helped define the goals for and contents of the book and provided some very useful reviews of early drafts of several of the chapters.

Thanks are due to Professor Leonard Kleinrock of UCLA. He read an early outline and several preliminary chapters and encouraged me to proceed. His two volume opus on queueing theory has been a great inspiration for me; it is an outstanding example of how technical writing should be done.

A number of people from the Hewlett-Packard Performance Technology Center supported my writing efforts. Philippe Benard has been of tremendous assistance. He helped conquer the dynamic interfaces between UNIX, FrameMaker, and *Mathematica*. He solved several difficult problems for me including discovering a method for importing *Mathematica* graphics into FrameMaker and coercing FrameMaker into producing a proper Table of Contents. Tom Milner became my UNIX advisor when Philippe moved to the Hewlett-Packard Cupertino facility. Jane Arteaga provided a number of graphics from Performance Technology Center documents in a format that could be imported into FrameMaker. Helen Fong advised me on RTEs, created a nice graphic for me, proofed several chapters, and checked out some of the *Mathematica* code. Jim Lewis read several drafts of the book, found some typos, made some excellent suggestions for changes, and ran most of the *Mathematica* code. Joe Wihnyk showed me how to force the FrameMaker HELP system to provide useful information. Paul Primmer, Richard Santos, and Mel Eelkema made suggestions about code profilers and SPT/iX. Mel also helped me describe the expert system facility of HP GlancePlus for MPE/iX. Rick Bowers proofed several chapters, made some helpful suggestions, and contributed a solution for an exercise. Jim Squires proofed several chapters, and made some excellent suggestions. Gerry Wade provided some insight into how collectors, software monitors, and diagnostic tools work. Sharon Riddle and Lisa Nelson provided some excellent graphics. Dave Gershon converted them to a format acceptable to FrameMaker. Tim Gross advised me on simulation and handled some ticklish UNIX problems. Norbert Vicente installed FrameMaker and Mathematica for me and customized my workstation. Dean Coggins helped me keep my workstation going.

Some Hewlett-Packard employees at other locations also provided support for the book. Frank Rowand and Brian Carroll from Cupertino commented on a draft of the book. John Graf from Sunnyvale counseled me on how to measure the CPU power of PCs. Peter Friedenbach, former Chairman of the Executive Steering Committee of the Transaction Processing Performance Council (TPC), advised me on the TPC benchmarks and provided me with the latest TPC benchmark results. Larry Gray from Fort Collins helped me understand the goals of the

Standard Performance Evaluation Corporation (SPEC) and the new SPEC benchmarks. Larry is very active in SPEC. He is a member of the Board of Directors, Chair of the SPEC Planning Committee, and a member of the SPEC Steering Committee. Dr. Bruce Spenner, the General Manager of Disk Memory at Boise, advised me on Hewlett-Packard I/O products. Randi Braunwalder from the same facility provided the specifications for specific products such as the 1.3- inch Kittyhawk drive.

Several people from outside Hewlett-Packard also made contributions. Jim Calaway, Manager of Systems Programming for the State of Utah, provided some of his own papers as well as some hard- to-find IBM manuals, and reviewed the manuscript for me. Dr. Barry Merrill from Merrill Consultants reviewed my comments on SMF and RMF. Pat Artis from Performance Associates, Inc. reviewed my comments on IBM I/O and provided me with the manuscript of his book, *MVS I/O Subsystems: Configuration Management and Performance Analysis,* McGraw-Hill, as well as his Ph. D. Dissertation. (His coauthor for the book is Gilbert E. Houtekamer.) Steve Samson from Candle Corporation gave me permission to quote from several of his papers and counseled me on the MVS operating system. Dr. Anil Sahai from Amdahl Corporation reviewed my discussion of IBM I/O devices and made suggestions for improvement. Yu-Ping Chen proofed several chapters. Sean Conley, Chris Markham, and Marilyn Gibbons from Frame Technology Technical Support provided extensive help in improving the appearance of the book. Marilyn Gibbons was especially helpful in getting the book into the exact format desired by my publisher. Brenda Feltham from Frame Technology answered my questions about the Microsoft Windows version of FrameMaker. The book was typeset using FrameMaker on a Hewlett-Packard workstation and on an IBM PC compatible running under Microsoft Windows. Thanks are due to Paul R. Robichaux and Carol Kaplan for making Sean, Chris, Marilyn, and Brenda available. Dr. T. Leo Lo of McDonnell Douglas reviewed Chapter 7 and made several excellent recommendations. Brad Horn and Ben Friedman from Wolfram Research provided outstanding advice on how to use *Mathematica* more effectively.

Thanks are due to Wolfram Research not only for asking Brad Horn and Ben Friedman to counsel me about *Mathematica* but also for providing me with *Mathematica* for my personal computer and for the HP 9000 computer that supported my workstation. The address of Wolfram Research is
Wolfram Research, Inc.
P. O. Box 6059
Champaign, Illinois 61821
Telephone: (217) 398-0700

Brian Miller, my production editor at Academic Press Boston did an excellent job in producing the book under a heavy time schedule. Finally, I would like to thank Jenifer Niles, my editor at Academic Press Professional, for her encouragement and support during the sometimes frustrating task of writing this book.

Reference

1. Martha L. Abell and James P. Braselton, *Mathematica by Example,* Academic Press, 1992.

2. Martha L. Abell and James P. Braselton, *The Mathematica Handbook,* Academic Press, 1992.

3. Nancy R. Blachman, *Mathematica: A Practical Approach*, Prentice-Hall, 1992.

4. Richard E. Crandall, *Mathematica for the Sciences*, Addison-Wesley, 1991.

5. Theodore Gray and Jerry Glynn, *Exploring Mathematics with Mathematica,* Addison-Wesley, 1991.

6. Leonard Kleinrock, *Queueing Systems, Volume I: Theory*, John Wiley, 1975.

7. Leonard Kleinrock, *Queueing Systems, Volume II: Computer Applications*, John Wiley, 1976.

8. Roman Maeder, *Programming in Mathematica,* Second Edition, Addison-Wesley, 1991.

9. Stan Wagon, *Mathematica in Action*, W. H. Freeman, 1991

10. Stephen Wolfram, *Mathematica: A System for Doing Mathematics by Computer, Second Edition*, Addison-Wesley, 1991.

Definitions and Notation

Digerati	**Digerati**, *n.pl.*, people highly skilled in the processing and manipulation of digital information; wealthy or scholarly techno-nerds. Definition by Tim Race
KB	**Kilobyte**. A memory size of $1024 = 2^{10}$ bytes.

GB	**Gigabyte.** A memory size of $1,073,741,824 = 2^{30}$ bytes.
MB	**Megabyte.** A memory size of $1,048,576 = 2^{20}$ bytes.
MFLOPS	Millions of floating point operations per second.
MHz	**Megahertz** or cyclic rate in millions of cycles per second.
ms	**Milliseconds.** One ms is 1/1000 seconds.
ns	**Nanoseconds.** One ns is 10^{-9} seconds.
RPM	Revolutions per minute. Used to specify the rotational speed of a disk drive.
SUT	System under test. A benchmarking abbreviation.
superscalar	A processor that issues multiple independent instructions per clock cycle.
TB	**Terabyte.** A memory size of $1,073,741,824 = 2^{40}$ bytes.
TFLOPS	**TeraFLOPS** or one million MFLOPS of floating point operations per second.

Chapter 1 Introduction

"I don't know what you mean by 'glory,'" Alice said. Humpty Dumpty smiled contemptuously. "Of course you don't—til I tell you. I meant 'there's a nice knock-down argument for you!'" "But 'glory' doesn't mean 'a nice knock-down argument,'" Alice objected. "When I use a word," Humpty Dumpty said, in a rather scornful tone, "it means just what I choose it to mean—neither more nor less." "The question is," said Alice, "whether you can make words mean so many different things." "The question is," said Humpty Dumpty, "which is to be master—that's all."

Lewis Carroll
Through The Looking Glass

A computer can never have too much memory or too fast a CPU.

Michael Doob
Notices of the AMS

1.1 Introduction

The word *performance* in computer performance means the same thing that performance means in other contexts, that is, it means "How well is the computer system doing the work it is supposed to do?" Thus it means the same thing for personal computers, workstations, minicomputers, midsize computers, mainframes, and supercomputers. Almost everyone has a personal computer but very few people think their PC is too fast. Most would like a more powerful model so that Microsoft Windows would come up faster and/or their spreadsheets would run faster and/or their word processor would perform better, etc. Of course a more powerful machine also costs more. I have a fairly powerful personal computer at home; I would be willing to pay up to $1500 to upgrade my machine if it would run *Mathematica* programs at least twice as fast. To me that represents good performance because I spend a lot of time running *Mathematica* programs and they run slower than any other programs I run. It is more difficult to decide what good or even acceptable performance is for a computer system used in business. It depends a great deal on what the computer is used for; we call the work the computer does the *workload.* For some applications, such as an airline reservation system, poor performance could cost an airline millions of dollars per day in lost revenue. Merrill has a chapter in his excellent book [Merrill 1984] called "Obtaining Agreement on Service Objectives." (By "service objectives" Merrill is referring to how well the computer executes the workload.) Merrill says

> There are three ways to set the goal value of a service objective: a measure of the user's subjective perception, management dictate, and guidance from others' experiences.
>
> Of course, the best method for setting the service objective goal value requires the most effort. Record the user's subjective perception of response and then correlate perception with internal response measures.

Merrill describes a case study that was used to set the goal for a CICS (Customer Information Control System, one of the most popular IBM mainframe application programs) system with 24 operators at one location. (IBM announced in September 1992 that CICS will be ported to IBM RS/6000 systems as well as to Hewlett-Packard HP 3000 and HP 9000 platforms.) For two weeks each of the 24 operators rated the response time at the end of each hour with the subjective ratings of Excellent, Good, Fair, Poor, or Rotten (the operators were not given any actual response times). After throwing out the outliers, the ratings were compared to the response time measurements from the CICS Performance Analyzer (an IBM CICS performance measurement tool). It was discovered that whenever over 93% of the CICS transactions completed in under 4 seconds, all operators rated the service as Excellent or Good. When the percentage dropped below 89% the operators rated the service as Poor or Rotten. Therefore, the service objective goal was set such that 90% of CICS transactions must complete in 4 seconds.

We will discuss the problem of determining acceptable performance in a business environment in more detail later in the chapter.

Since acceptable computer performance is important for most businesses we have an important sounding phrase for describing the management of computer performance—it is called *performance management* or *capacity management*.

P*erformance management* is an umbrella term to include most operations and resource management aspects of computer performance. There are various ways of breaking performance management down into components. At the Hewlett-Packard Performance Technology Center we segment performance management as shown in Figure 1.1.

We believe there is a core area consisting of common access routines that provide access to performance metrics regardless of the operating system platform. Each quadrant of the figure is concerned with a different aspect of performance management.

Application optimization helps to answer questions such as "Why is the program I use so slow?" Tools such as profilers can be used to improve the performance of application code, and other tools can be used to improve the efficiency of operating systems.

Figure 1.1. Performance Management

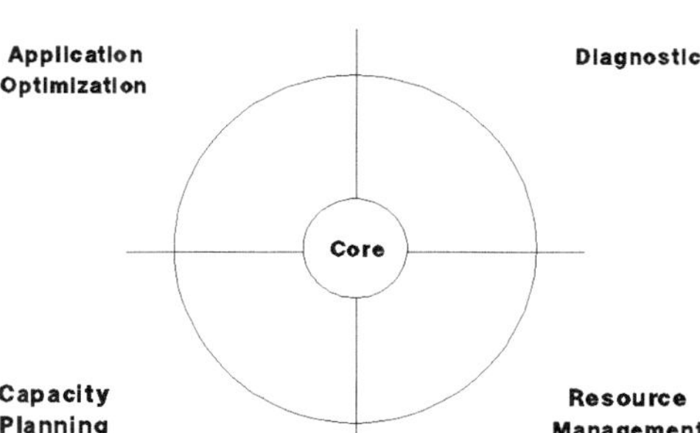

A profiler is an important tool for improving the efficiency of a program by indicating which sections of the code are used the most. A widely held rule of thumb is that a program spends 90% of its execution time in only 10% of the code. Obviously the most executed parts of the code are where code improvement efforts should be concentrated. In his classic paper [Knuth 1971] Knuth claimed in part, "We also found that less than 4 percent of a program generally accounts for more than half of its running time."

There is no sharp line between application optimization and system tuning.

Diagnosis deals with the determination of the causes of performance problems, such as degraded response time or unacceptable fluctuations in throughput. A diagnostic tool could help to answer questions such as "Why does the response time get so bad every afternoon at 2:30?" To answer questions such as this one, we must determine if there is a shortage of resources such as main memory, disk drives, CPU cycles, etc., or the system is out of tune or needs to be rescheduled. Whatever the problem, it must be determined before a solution can be obtained.

Resource management concerns include scheduling of the usage of existing resources in an optimal manner, system tuning, service level agreements, and load balancing. Thus resource management could answer the question "What is the best time to do the daily system backup?" We will discuss service level agreements later. Efficient installations balance loads across devices, CPUs, and systems and attempt to schedule resource intensive applications for off hours.

Capacity planning is more of a long-term activity than the other parts of performance management. The purpose of capacity planning is to provide an acceptable level of computer service to the organization while responding to workload demands generated by business requirements. Thus capacity planning might help to answer a question such as "Can I add 75 more users to my system?" Effective capacity planning requires an understanding of the sometimes conflicting relationships between business requirements, computer workload, computer capacity, and the service or responsiveness required by users.

These subcategories of performance management are not absolute—there is a fuzziness at the boundaries and the names change with time. At one time all aspects of it were called computer performance evaluation, abbreviated CPE, and the emphasis was upon measurement. This explains the name *Computer Measurement Group* for the oldest professional organization dealing with computer performance issues. (We discuss this important organization later in the chapter.)

In this book we emphasize the capacity planning part of computer performance management. That is, we are mainly concerned not with day-to-day activities but rather with what will happen six months or more from today. Note that most of the techniques that are used in capacity planning are also useful for application optimization. For example, Boyse and Warn [Boyse and Warn 1975] show how queueing models can be used to decide whether an optimizing compiler should be purchased and to decide how to tune the system by setting the multiprogramming level.

The reasons often heard for not having a program of performance management in place but rather acting in a reactive manner, that is, taking a "seat of the pants" approach, include:

1. We are too busy fighting fires.
2. We don't have the budget.
3. Computers are so cheap we don't have to plan.

The most common reason an installation has to fight fires is that the installation does not plan ahead. Lack of planning causes crises to develop, that is, starts the fires. For example, if there is advance knowledge that a special application will require more computer resources for completion than are currently available, then arrangements can be made to procure the required capacity before they are required. It is not knowing what the requirements are that can lead to panic.

Investing in performance management saves money. Having limited resources is thus a compelling reason to do more planning rather than less. It doesn't require a large effort to avoid many really catastrophic problems.

With regard to the last item there are some who ask: "Since computer systems are getting cheaper and more powerful every day, why don't we solve any capacity shortage problem by simply adding more equipment? Wouldn't this be less expensive than using the time of highly paid staff people to do a detailed systems analysis for the best upgrade solution?" There are at least three problems with this solution. The first is that, even though the cost of computing power is declining, most companies are spending more on computing every year because they are developing new applications. Many of these new applications make sense only because computer systems are declining in cost. Thus the computing budget is increasing and the executives in charge of this resource must compete with other executives for funds. A good performance management effort makes it easier to justify expenditures for computing resources.

Another advantage of a good performance management program is that it makes the procurement of upgrades more cost effective (this will help get the required budget, too).

A major use of performance management is to prevent a sudden crisis in computer capacity. Without it there may be a performance crisis in a major application, which could cost the company dearly.

In organizing performance management we must remember that hardware is not the only resource involved in computer performance. Other factors include how well the computer systems are tuned, the efficiency of the software, the operating system chosen, and priority assignments.

It is true that the performance of a computer system does depend on hardware resources including

1. the speed of the CPU or CPUs
2. the size and speed of main memory
3. the size and speed of the memory cache between the CPU and main memory
4. the size and speed of disk memory

5. the number and speed of I/O channels and the size as well as the speed of disk cache (on disk controllers or in main memory)
6. tape memory
7. the speed of the communication lines connecting the terminals or workstations to the computer system.

However, as we mentioned earlier, the performance also depends on

1. the operating system that is chosen
2. how well the system is tuned
3. how efficiently locks on data bases are used
4. the efficiency of the application software, and
5. the scheduling and priority assignments.

This list is incomplete but provides some idea of the scope of computer performance. We discuss the components of computer performance in more detail in Chapter 2.

1.2 Capacity Planning

Capacity planning is the most challenging of the four aspects of performance management. We consider some of the difficulties in doing effective capacity planning next.

Difficulty of Predicting Future Workloads

To do this successfully, the capacity planner must be aware of all company business plans that affect the computer installation under study. Thus, if four months from now 100 more users will be assigned to the installation, it is important to plan for this increase in workload now.

Difficulty in Predicting Changes in Technology

According to Hennessy and Patterson [Hennessy and Patterson 1990] the performance growth rate for supercomputers, minicomputers, and mainframes has recently been about 20% per year while for microcomputers it has been about 35% per year. However, for computers that use RISC technology the growth rate has been almost 100% per year! (RISC means "reduced instruction set computers" as compared to the traditional CISC or "complex instruction set computers.") Similar rates of improvement are being made in main memory technology. Unfortunately, the improvement rate for I/O devices lags behind those for other technologies. These changes must be kept in mind when planning future upgrades.

In spite of the difficulties inherent in capacity planning, many progressive companies have successful capacity planning programs. For the story of how the M&G Group PLC of England successfully set up capacity planning at an IBM mainframe installation see the interesting article [Claridge 1992]. There are four parts of a successful program:

1. understanding the current business requirements and user's performance requirements
2. prediction of future workload
3. an evaluation of future configurations
4. an ongoing management process.

We consider each of these aspects in turn.

1.2.1 Understanding the Current Environment

Some computer installations are managed in a completely reactive manner. No problem is predicted, planned for, or corrected until it becomes a crisis. We believe that an orderly, planned, approach to every endeavor should be taken to avoid being "crisis or event driven." To be successful in managing our computer resources, we must take our responsibility for the orderly operation of our computer facilities seriously, that is, we must become more proactive.

To become proactive, we must understand the current business requirements of the organization, understand our current workload and the performance of our computer systems in processing that workload, and understand the user's service expectations. In short, we must understand our current situation before we can plan for the future.

As part of this effort the workload must be carefully defined in terms that are meaningful both to the end user and the capacity planner. For example, a workload class might be interactive order entry. For this class the workload could be described from the point of view of the users as orders processed per day. The capacity planner must convert this description into computer resources needed per order entered; that is, into CPU seconds per transaction, I/Os required per transaction, memory required, etc.

Devising a measurement strategy for assessing the actual performance and utilization of a computer system and its components is an important part of capacity planning. We must obtain the capability for measuring performance and for storing the performance data for later reference, that is, we must have measurement tools and a performance database. The kind of program that collects system resource consumption data on a continuous basis is called a "software monitor" and the performance data files produced by a monitor are often called "log files." For example, the Hewlett-Packard performance tool HP LaserRX has a monitor called SCOPE that collects performance information and stores it for later use in log files. If you have an IBM mainframe running under the MVS operating system, the monitor most commonly used is the IBM Resource Measurement Facility (RMF). From the performance information that has been captured we can determine what our current service levels are, that is, how well we are serving our customers. Other tools exist that make it easy for us to analyze the performance data and present it in meaningful ways to users and management. An example is shown in Figure 1.2, which was provided by the Hewlett-Packard UNIX performance measurement tool HP LaserRX/UX. HP LaserRX/UX software lets you display and analyze collected data from one or more HP-UX based systems. This figure shows how you can examine a graph called "Global Bottlenecks," which does not directly indicate bottlenecks but does show the major resource utilization at the global level, view CPU system utilization at the global level, and then make a more detailed inspection at the application and process level. Thus we examine our system first from an overall point of view and then hone in on more detailed information. We discuss performance tools in more detail later in this chapter.

Once we have determined how well our current computer systems are supporting the major applications we need to set performance objectives.

1.2.1.1 Performance Measures

The two most common performance measures for interactive processing are average *response time* and average *throughput*. The first of these measures is the delay the user experiences between the instant a request for service from the computer system is made and when the computer responds. The average throughput is a measure of how fast the computer system is processing the work. The precise value of an individual response time is the elapsed time from the instant the user hits the enter key until the instant the corresponding reply begins to appear on the monitor of the workstation or terminal. Performance analysts often call the response time we defined as "time to first response" to distinguish it from "time to prompt." (The latter measures the interval from the instant the user hits the enter key until the entire response has appeared at the terminal and a prompt symbol appears.) If, during an interval of time, n responses have been received of lengths $l_1, l_2, \ldots, l_n$, then the average response time R is defined the same way an instructor calculates the average grade of an exam: by adding up all the grades and dividing by the number of students. Thus $R = (l_1 + l_2 + \ldots + l_n)/n$. Since a great deal of variability in response time disturbs users, we sometimes compute measures of the variability as well, but we shall not go into this aspect of response time here.

Figure 1.2. HP LaserRX/UX Example

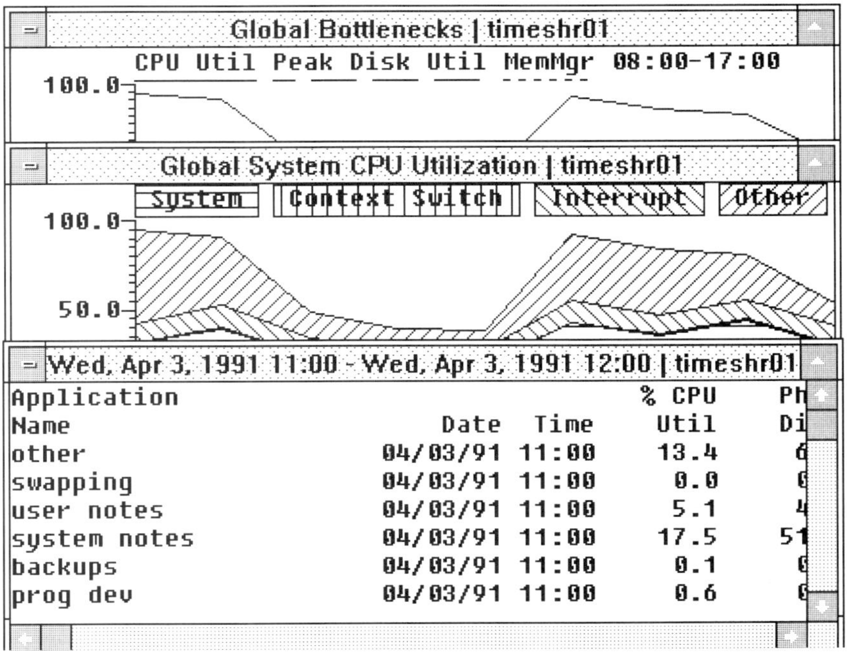

Global Bottlenecks | timeshr01

CPU Util Peak Disk Util MemMgr 08:00-17:00

100.0

Global System CPU Utilization | timeshr01

System Context Switch Interrupt Other

100.0

50.0

Wed, Apr 3, 1991 11:00 - Wed, Apr 3, 1991 12:00 | timeshr01

Application Name	Date	Time	% CPU Util	Ph Di
other	04/03/91	11:00	13.4	6
swapping	04/03/91	11:00	0.0	0
user notes	04/03/91	11:00	5.1	4
system notes	04/03/91	11:00	17.5	51
backups	04/03/91	11:00	0.1	0
prog dev	04/03/91	11:00	0.6	0

Another response time performance parameter is the pth *percentile* of response time, which is defined to be the value of response time such that p percent of the observed values do not exceed it. Thus the 90th percentile value of response time is exceeded by only 10 percent of the observed values. This means that 1 out of 10 values will exceed the 90th percentile value. It is part of the folklore of capacity planning that the perceived value of the average response time experienced is the 90th percentile value of the actual value. If the response time has an exponential distribution (a common occurrence) then the 90th percentile value is 2.3 times the average value. Thus, if a user has experienced a long sequence of exponentially distributed response times with an average value of 2 seconds, the user will perceive an average response time of 4.6 seconds! The reason for this is as follows: Although only 1 out of 10 response times exceeds 4.6 seconds, these long response times make a bigger impression on the memory than the 9 out of 10 that are smaller. We all seem to remember bad news better than good news! (Maybe that's why most of the news in the daily paper seems to be bad news.)

The average *throughput* is the average rate at which jobs are completed in an interval of time, that is, the number of jobs or transactions completed divided by the time in which they were completed. Thus, for an order-entry application, the throughput might be measured in units of number of orders entered per hour, that is, orders per hour. The average throughput is of more interest to management than to the end user at the terminal; it is not sensed by the users as response time is, but it is important as a measure of productivity. It measures whether or not the work is getting done on time. Thus, if Short Shingles receives 4,000 orders per day but the measured throughput of their computer system is only 3,500 order-entry applications per day, then the orders are not being processed on time. Either the computer system is not keeping up, there are not enough order-entry personnel to handle all the work, or some other problem exists. Something needs to be done!

The primary performance measures for batch processing are average job turnaround time and average throughput. Another important performance measure is completion of the batch job in the "batch window" for installations that have an important batch job that must be completed within a "window." The window of such a batch job is the time period in which it must be started and completed. The payroll is such an application. It cannot be started until the work records of the employees are available and must be completed by a fixed time or there will be a lot of disgruntled employees. An individual job turnaround time is the interval between the instant a batch program (job) is read into the computer system and the instant that the program completes execution. Thus a batch system processing bills to customers for services rendered might have a turnaround time of 12 minutes and a throughput of three jobs per hour.

Another performance measure of interest to user departments is the availability of the computer system. This is defined as the percentage of scheduled computer system time in which the system is actually available to users to do useful work. The system can fail to be available because of hardware failures, software failures, or by allowing preventive maintenance to be scheduled during normal operating hours.

1.2.2 Setting Performance Objectives

From the management perspective, one of the key aspects of capacity planning is setting the performance objectives. (You can't tell whether or not you are meeting your objectives if you do not have any.) This involves negotiation between user groups and the computer center management or information systems (IS) group.

One technique that has great potential is a *service level agreement* between IS and the user departments.

Service Level Agreements

A service level agreement is a contract between the provider of the service (IS, MIS, DP, or whatever the provider is called) and the end users that establishes mutual responsibilities for the service to be provided. The computer installation management is responsible for providing the agreed-upon service (response time, availability, throughput, etc.) as well as the measurement and reporting of the service provided. To receive the contracted service, the end users must agree to certain volumes and mix of work. For example, the end user department must agree to provide the input for a batch job by a certain time, say, 10 a.m. The department might also agree to limit the number of terminals or workstations active at any one time to 350, and that the load level of online transactions from 2 p.m. to 5 p.m. would not exceed 50 transactions per second. If these and other stipulations are exceeded or not met, then the promised service cannot be guaranteed.

Several useful processes are provided by service level agreements. Capacity planners are provided with a periodic review process for examining current workload levels and planning future levels. User management has an opportunity to review the service levels being provided and for making changes to the service objectives if this proves desirable. The installation management is provided with a process for planning and justifying future resources, services, and direction.

Ideally, service level objectives are established as a result of the business objectives. The purpose of the service level objectives is to optimize investment and revenue opportunities. Objectives are usually stated in terms of a range or an average plus a percentile value, such as average online response time between

0.25 and 1.5 seconds during the peak period of the day, or as an average of 1.25 seconds with a 95th percentile response time of 3.75 seconds at all times. The objectives usually vary by time of day, day of the week, day of the month, type of work, and by other factors, such as a holiday season, that can impact performance. Service level objectives are usually established for online response time, batch turnaround time, availability requirements for resources and workloads, backup and recovery resources and procedures, and disaster plans.

McBride [McBride 1990] discusses some of the procedural issues in setting up an SLA as follows:

> Before MIS goes running off to talk to users about establishing SLAs, they need to know the current DP environment in terms of available hardware and software, what the current demands are on the hardware/software resource set, what the remaining capacity is of the resource set, and they need to know the current service levels.
>
> Once this information has been captured and understood within the context of the data processing organization, users representing the various major applications supported by MIS should be queried as to what their expectations are for DP service. Typically, users will be able to respond with qualitative, rather than quantitative, answers regarding their current and desired perceptions of service levels. Rather than saying "95th percentile response times should be less than or equal to X," they'll respond with, "I need to be able to keep my data entry people focused on their work, and I need to be able to handle my current claim load without falling behind."
>
> It is MIS's responsibility to take this qualitative information and quantify it in order to relate to actual computer resource consumption. This will comprise a starting point from which actual SLAs can be developed. By working with users to determine what their minimum service levels are, as well as determining how the user's demand on DP resources will change as the company grows, MIS can be prepared to predict when additional resources will be needed to continue to meet the users demands. Alternatively, MIS will be able to predict when service levels will no longer be met and what the result-

ing service levels will be without the acquisition of additional resources.

One of the major advantages of the use of SLAs is that it gets a dialog going between the user departments and the computer installation management. This two-way communication helps system management understand the needs of their users and it helps the users understand the problems IS management has in providing the level of service desired by the users. As Backman [Backman 1990] says about SLA benefits:

> The expectations of both the supplier and the consumer are set. Both sides are in agreement on the service and the associated criteria defined. This is the main tangible benefit of using SLAs.
>
> The intangible benefits, however, provide much to the parties as well. The transition from a reactionary fire fighting methodology of performance management to one of a proactive nature will be apparent if the SLA is followed and supported. Just think how you will feel if all those "system surprises" have been eliminated, allowing you to think about the future. The SLA method provides a framework for organizational cooperation. The days of frantically running around juggling batch schedules and moving applications from machine to machine are eliminated if the SLA has been properly defined and adhered to.
>
> Also, capacity planning becomes a normal, scheduled event. Regular capacity planning reports will save money in the long run since the output of the capacity plan will be factored into future SLAs over time, allowing for the planned increases in volume to be used in the projection of future hardware purchases.

Miller in his article [Miller 1987] on service level agreements claims the elements that need to be structured for a successful service level agreement are as follows:

1. Identify the *parties* to the agreement.
2. Describe the *service* to be provided.
3. Specify the *volume* of demand for service over time.

4. Define the *timeliness* requirements for the service.
5. Discuss the *accuracy* requirements.
6. Specify the *availability* of the service required.
7. Define the *reliability* of the service provided.
8. Identify the limitations to the service that are acceptable.
9. Quantify the *compensation* for providing the service.
10. Describe the *measurement* procedures to be used.
11. Set the date for *renegotiation* of the agreement.

Miller also provides a proposed general format for service level agreements and an excellent service level agreement checklist.

If service level agreements are to work well, there must be cooperation and understanding between the users and the suppliers of the information systems. Vanvick in his interesting paper [Vanvick 1992] provides a quiz to be taken by IS managers and user managers to help them understand each other. He recommends that IS respondents with a poor score get one week in a user re-education camp where acronyms are prohibited. User managers get one week in an IS re-education camp where acronyms are the only means of communication.

Another tool that is often used in conjunction with service level agreements is *chargeback* to the consumer of computer resources.

Chargeback

There are those who believe that a service level agreement is a carrot to encourage user interest in performance management while chargeback is the stick. That is, if users are charged for the IS resources they receive, they will be less likely to make unrealistic performance demands. In addition users can sometimes be persuaded to shift some of their processing to times other than the peak period of the day by offering them lower rates.

Not all installations use chargeback but some types of installations have no choice. For example, universities usually have a chargeback system to prevent students from using excessive amounts of IS resources. Students usually have job identification numbers; a limited amount of computing is allowed for each number.

According to Freimayer [Freimayer 1988] benefits of a chargeback system include the following:

1. Performs budget and usage forecasting.
2. Promotes cost effective computer resource utilization.
3. Encourages user education concerning the cost associated with individual data processing usage.
4. Helps identify data processing overhead costs.
5. Identifies redundant or unnecessary processing.
6. Provides a method for reporting data processing services rendered.
7. Increases data center and user accountability.

These seem to be real benefits but, like most things in this world, they are not obtained without effort. The problems with chargeback systems are always more political than technical, especially if a chargeback system is just being implemented. Most operating systems provide the facilities for collecting the information needed for a chargeback program and commercial software is available for implementing chargeback. The difficulties are in deciding the goals of a program and implementing the program in a way that will be acceptable to the users and to upper management.

The key to implementing a chargeback program is to treat it as a project to be managed just as any other project is managed. This means that the goals of the project must be clearly formulated. Some typical goals are:

1. Recover the full cost to IS for the service provided.
2. Encourage users to take actions that will improve performance, such as performing low priority processing at off-peak times, deleting obsolete data from disk storage, and moving some processing such as word processing or spreadsheets to PCs or workstations.
3. Discourage users from demanding unreasonable service levels.

Part of the implementation project is to ensure that the users understand and feel comfortable with the goals of the chargeback system that is to be implemented. It is important that the system be perceived as being fair. Only then should the actual chargeback system be designed and implemented. Two important parts of the project are: (1) to get executive level management approval and (2) to verify with the accounting department that the accounting practices used in the plan meet company standards. Then the chargeback algorithms can be designed and put into effect.

Some of the components that are often combined in a billing algorithm include:

1. CPU time
2. disk I/O
3. disk space used (quantity and duration)
4. tape I/O
5. connect time
6. network costs
7. paging rate
8. lines printed
9. amount of storage used real/virtual).

Factors that may affect the billing rates of the above resources include:

1. job class
2. job priority surcharges
3. day shift (premium)
4. evening shift (discount).

As an example of how a charge might be levied, suppose that the CPU cost per month for a certain computer is \$100,000 and that the number of hours of CPU time used in October was 200. Then the CPU billing rate for October would be \$100,000/200 = \$500 per hour, assuming there were no premium charges. If Group A used 10 hours of CPU time in October, the group would be charged \$5,000 for CPU time plus charges for other items that were billable such as the disk I/O, lines printed, and amount of storage used.

Standard costing is another method of chargeback that can be used for mature systems, that is, systems that have been in use long enough that IS knows how much of each computer resource is needed, on the average, to process one of the standard units, also called a business work unit (BWU) or natural forecasting unit (NFU). An example for a travel agency might be a booking of an airline flight. For a bank it might be the processing of a monthly checking account for a private (not business) customer. A BWU for a catalog service that takes most orders by 800 number phone calls could be phone orders processed.

Other questions that must be answered as part of the implementation project include:

1. What reports must be part of the chargeback process and who receives them?

2. How are disagreements about charges negotiated?

3. When is the chargeback system reviewed?

4. When is the chargeback system renegotiated?

A chargeback system works best when combined with a service level agreement so both can be negotiated at the same time.

Schrier [Schrier 1992] described how the City of Seattle developed a chargeback system for a data communications network.

Not everyone agrees that chargeback is a good idea; especially when disgruntled users can buy their own PCs or workstations. The article by Butler [Butler 1992] contains interviews with a number of movers and shakers as well as a discussion of the tools available for chargeback. The subtitle of the article is, "Users, IS disagree on chargeback merit for cost control in downsized environment." The abstract is:

> Chargeback originated as a means of allocating IS costs to their true users. This was a lot simpler when the mainframe did all the computing. Proponents argue that chargeback is still needed in a networked environment. At Lawrence Berkeley Lab, however, support for chargeback has eroded as the role of central computers has diminished.

Clearly, sweeping changes are occurring in the computing environment.

Software Performance Engineering (SPE)

Software performance engineering is another relatively new discipline. It has become more evident in recent years that the proper time to think about the performance of a new application is while it is being designed and coded rather than after it has been coded and tested for functional correctness. There are many "war stories" in circulation about systems designed using the old style "fix-it-later" approach based on the following beliefs:

1. Performance problems are rare.

2. Hardware is fast and inexpensive.

3. It is too expensive to build high performance software.

4. Tuning can be done later.

5. Efficiency implies tricky code.

The fix-it-later approach assumes that it is not necessary to be concerned with performance considerations until after application development is complete. Proponents of this approach believe that any performance problems that appear after the system goes into production can be fixed at that time. The preceding list of reasons is given to support this view. We comment on each of the reasons in the following paragraphs.

It may have been true at one time that performance problems are rare but very few people would agree with that assessment today. The main reason that performance problems are less rare is that systems have gotten much more complicated, which makes it more difficult to spot potential performance problems.

It is true that new hardware is faster and less expensive every year. However, it is easy to design a system that can overwhelm any hardware that can be thrown at it. In other cases a hardware solution to a poor design is possible but at a prohibitive cost; hardware is never free!

The performance improvement that can be achieved by tuning is very limited. To make major improvements, it is usually necessary to make major design changes. These are hard to implement once an application is in production.

Smith [Smith 1991] gives an example of an electronic funds transfer system that was developed by a bank to transfer as much as 100 billion dollars per night. Fortunately the original design was checked by performance analysis personnel who showed that the system could not transfer more than 50 billion per night. If the original system had been developed, the bank would have lost the interest on 50 billion dollars every night until the system was fixed.

It is a myth that only tricky code can be efficient. Tricky code is sometimes developed in an effort to improve the performance of a system after it is developed. Even if it succeeds in improving the performance, the tricky code is difficult to maintain. It is much better to design the good performance into the software from the beginning without resorting to nonstandard code.

A new software discipline, Software Performance Engineering, abbreviated SPE, has been developed in the last few years to help software developers ensure that application software will meet performance goals at the end of the development cycle. The standard book on SPE is [Smith 1991]. Smith says, in the opening paragraph:

> Software Performance Engineering (SPE) is a method for constructing software systems to meet performance objectives. The process begins early in the software lifecycle and uses quantitative methods to identify satisfactory designs and to eliminate those that are likely to have unacceptable performance, before developers invest significant time in implementation. SPE continues through the detailed design, coding, and testing stages to predict and manage the performance of the evolving software and to monitor and report actual performance against specifications and predictions. SPE methods cover performance data collection, quantitative analysis techniques, prediction strategies, management of uncertainties, data presentation and tracking, model verification and validation, critical success factors, and performance design principles.

The basic principle of SPE is that service level objectives are set during the application specification phase of development and are designed in as the functionality of the application is specified and detailed design begins. Furthermore, resource requirements to achieve the desired service levels are also part of the development process.

One of the key techniques of SPE is the *performance walkthrough.* It is performed early in the software development cycle, in the requirements analysis phase, as soon as a general idea of system functions is available. The main part of the meeting is a walkthrough of the major system functions to determine whether or not the basic design can provide the desired performance with the anticipated volume of work and the envisioned hardware platform. An example of how this might work is provided by Bailey [Bailey 1991]. A database transaction processing system was being designed that was required to process 14 transactions per second during the peak period of the day. Each transaction required the execution of approximately 1 million computer instructions on the proposed computer. Since the computer could process far in excess of 14 million instructions per second, it appeared there would be no performance problems. However, closer inspection revealed that the proposed computer was a multiprocessor with four CPUs and that the database system was single threaded, that is, to achieve the required performance each processor would need the capability of processing 14 million instructions per second! Since a single CPU could not deliver the required CPU cycles the project was delayed until the database system was modified to allow multithreading operations, that is, so that four transactions could be executed simultaneously. When the database system was upgraded the project went forward and was very successful. Without the walkthrough the system would have been developed prematurely.

I believe that a good performance walkthrough could have prevented many, if not most, of the performance disasters that have occurred. However, Murphy's law must be repealed before we can be certain of the efficacy of performance walkthroughs. Of course the performance walkthrough is just the beginning of the SPE activity in a software development cycle, but a very important part.

Organizations that have adopted SPE claim that they need to spend very little time tuning their applications after they go into the production phase, have fewer unpleasant surprises just before putting their applications into production, and have a much better idea of what hardware resources will be needed to support their applications in the future. Application development done using SPE also requires less software maintenance, less emergency hardware procurement, and more efficient application development. These are strong claims, as one would expect from advocates, but SPE seems to be the wave of the future.

Howard in his interesting paper [Howard 1992a] points out that serious political questions can arise in implementing SPE. Howard says:

> SPE ensures that application development not only satisfies functional requirements, but also performance requirements.
>
> There is a problem that hinders the use of SPE for many shops, however. It is a political barrier between the application development group and other groups that have a vested interest in performance. This wall keeps internal departments from communicating information that can effectively increase the performance of software systems, and therefore decrease overall MIS operating cost.
>
> Lack of communication and cooperation is the greatest danger. This allows issues to slip away without being resolved. MIS and the corporation can pay dearly for system inefficiencies, and sometimes do not even know it.
>
> A commitment from management to improve communications is important. Establishing a common goal of software development—the success of the corporation—is also critical to achieving staff support. Finally, the use of performance anal-

> ysis tools can identify system problems while eliminating finger pointing.

Howard gives several real examples, without the names of the corporations involved, in which major software projects failed because of performance problems. He provides a list of representative performance management products with a description of what they do. He quotes from a number of experts and from several managers of successful projects who indicate why they were successful. It all comes down to the subtitle of Howard's paper, "To balance program performance and function, users, developers must share business goals."

Howard [Howard 1992b] amplifies some of his remarks in [Howard 1992a] and provides some helpful suggestions on selling SPE to application developers.

Never make forecasts; especially about the future.
Samuel Goldwyn

1.2.3 Prediction of Future Workload

To plan for the future we must, of course, be able to make a prediction of future workload. Without this prediction we cannot evaluate future configurations. One of the major goals of capacity planning is to be able to install upgrades in hardware and software on a timely basis to avoid the "big surprise" of the sudden discovery of a gross lack of system capacity. To avoid a sudden failure, it is necessary to predict future workload. Of course, predicting future workload is important for all timely upgrades.

It is impossible to make accurate forecasts without knowing the future business plans of the company. Thus the capacity planner must also be a business analyst; that is, must be familiar with the kind of business his or her enterprise does, such as banking, electronics manufacturing, etc., as well as the impact on computer system requirements because of particular business plans such as mergers, acquisitions, sales drives, etc. For example, if a capacity planner works for a bank and discovers that a marketing plan to get more customers to open checking accounts is being implemented, the planner must know what the impact of this sales plan will be on computer resource usage. Thus the capacity planner needs to know the amount of CPU time, disk space, etc., required for each checking account as well as the expected number of new checking accounts in order to predict the impact upon computer resource usage.

In addition to user input, capacity planners should know how to use statistical forecasting techniques including visual trending and time series regression models. We discuss these techniques briefly later in this chapter in the section on "statistical projection." More material about statistical projection techniques is provided in Chapter 7.

1.2.4 Evaluation of Future Configurations

To avoid shortages of computer capacity it is necessary to predict how the current system will perform with the predicted workload so it can be determined when upgrades to the system are necessary. The discipline necessary for making such predictions is *modeling*. For successful capacity planning it is also necessary to make performance evaluations of possible computer system configurations with the projected workload. Thus, this is another capacity planning function that requires modeling technology. As we show in Figure 1.3 there is a spectrum of modeling techniques available for performance prediction including:

1. rules of thumb
2. back-of-the-envelope calculations
3. statistical forecasting
4. analytical queueing theory modeling
5. simulation modeling
6. benchmarking.

Figure 1.3. Spectrum of Modeling Techniques

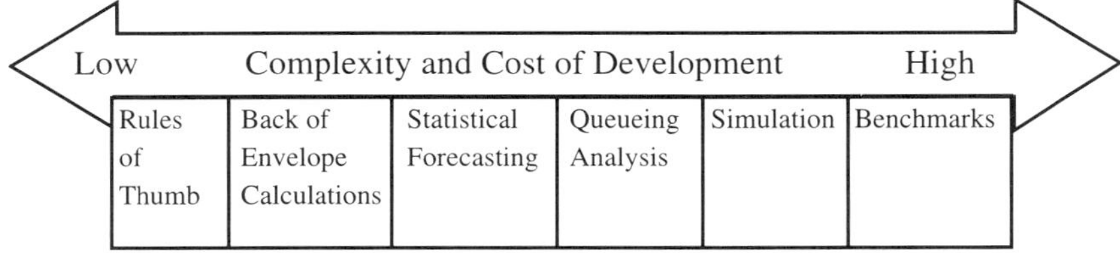

The techniques increase in complexity and cost of development from left to right in Figure 1.3 (top to bottom in the preceding list). Thus the application of rules of thumb is relatively straightforward and has little cost in time and effort. By contrast constructing and running a benchmark that faithfully represents the workload of the installation is very expensive and time consuming. It is not necessarily true that a more complex modeling technique leads to greater modeling accuracy. In particular, although benchmarking is the most difficult technique to apply, it is sometimes less accurate than analytical queueing theory modeling. The reason for this is the extreme difficulty of constructing a benchmark that faithfully models the actual workload. We discuss each of these modeling techniques briefly in this chapter. Some of them, such as analytic queueing theory modeling, will require an entire chapter of this book to explain adequately.

1.2.4.1 Rules of Thumb

Rules of thumb are guidelines that have developed over the years in a number of ways. Some of them are communicated by computer manufacturers to their customers and some are developed by computer users as a result of their experience. Every computer installation has developed some of its own rules of thumb from observing what works and what doesn't. Zimmer [Zimmer 1990] provides a number of rules of thumb including the load guidelines for data communication systems given in Table 1.1. If an installation does not have reliable statistics for estimating the load on a proposed data communication system, this table could be used. For example, if the system is to support 10 people performing data entry, 5 people doing inquiries, and 20 people with word processing activities, then the system must have the capability of supporting 10,000 data entry transactions, 1500 inquiry transactions, and 2000 word processing transactions per day.

The following performance rules of thumb have been developed by Hewlett-Packard performance specialists for HP 3000 computers running the MPE/iX operating system:

1. Memory manager CPU utilization should not exceed 8%.

2. Overall page fault rate should not exceed 30 per second. (We discuss page faults in Chapter 2.)

3. The time the CPU is paused for disk should not exceed 25%.

4. The utilization level for each disk should not exceed 80%.

There are different rules of thumb for Hewlett-Packard computer systems running under the HP-UX operating system. Other computer manufacturers have similar rules of thumb.

Table 1.1. Guidelines

Application	Typical Complexity	Trans/Term/ Person/Day
Data Entry	Simple	1,000
Inquiry	Medium	300
Update/ Inquiry	Complex	500
Personal Computer	Complex	100
Word Processing	Complex	100

Rosenberg [Rosenberg 1991] provides some general rules of thumb (which he attributes to his mentor, a senior systems programmer) such as:

1. There are only three components to any computer system-CPU, I/O, and memory.

Rosenberg says that if we want to analyze something not on this list, such as expanded memory on an IBM mainframe or on a personal computer, we can analyze it in terms of its effect on CPU, I/O, and memory.

He also provides a three-part rule of thumb for computer performance diagnosis that is valid for any computer system from a PC to a supercomputer:

1. If the CPU is at 100% utilization or less and the required work is being completed on time, everything is okay for now (but always remember, tomorrow is another day).

2. If the CPU is at 100% busy, and all work is not completed, you have a problem. Begin looking at the CPU resource.

3. If the CPU is not 100% busy, and all work is not being completed, a problem also exists and the I/O and memory subsystems should be investigated.

Rules of thumb are often used in conjunction with other modeling techniques as we will show later. As valuable as rules of thumb are, one must use caution in applying them because a particular rule may not apply to the system under consideration. For example, many of the rules of thumb given in [Zimmer 1990] are operating system dependent or hardware dependent; that is, may only be valid for systems using the IBM MVS operating system or for Tandem computer systems, etc.

Samson in his delightful paper [Samson 1988] points out that some rules of thumb are of doubtful authenticity. These include the following:

1. There is a knee in the curve.

2. Keep device utilization below 33%.

3. Keep path utilization below 30%.

4. Keep CPU utilization below ??%.

Figure 1.4. Queueing Time vs Utilization for M/M/1 System

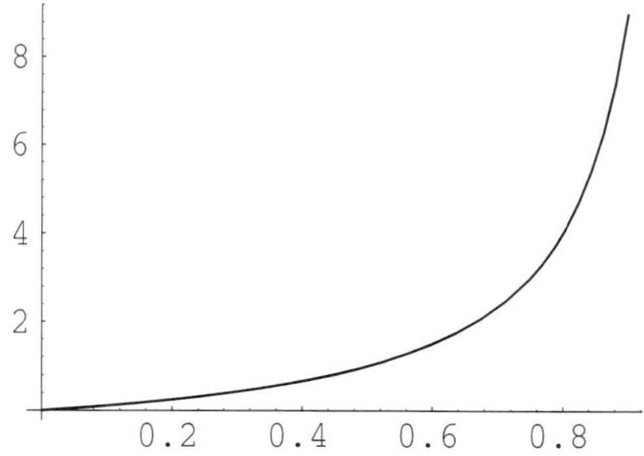

To understand these questionable rules of thumb you need to know about the curve of queueing time versus utilization for the simple M/M/1 queueing system. The M/M/1 designation means there is one service center with one server; this server provides exponentially distributed service. The M/M/1 system is an open system with customers arriving at the service center in a pattern such that the time between the arrival of consecutive customers has an exponential distribution. The curve of queueing time versus server utilization is smooth with a vertical asymptote at a utilization of 1. This curve is shown Figure 1.4. If we let S represent the average service time, that is, the time it takes the server to provide service to one customer, on the average, and U the server utilization, then the average queueing time for the M/M/1 queueing system is given by

$$U \times \frac{S}{1-U}.$$

Figure 1.5. A Mythical Curve

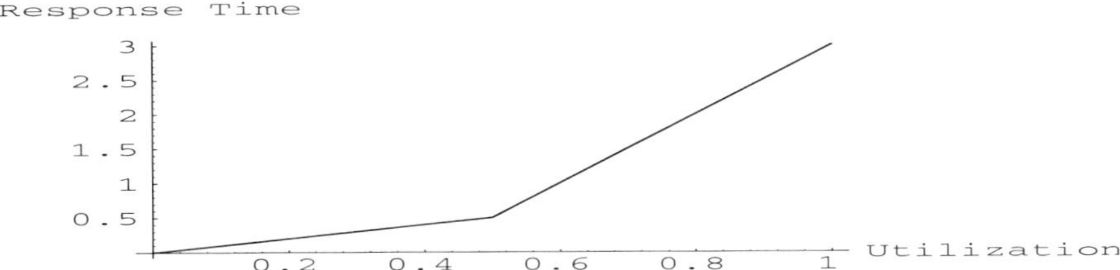

With regard to the first questionable rule of thumb (There is a knee in the curve), many performance analysts believe that, if response time or queueing time is plotted versus load on the system or device, then, at a magic value of load, the curve turns up sharply. This point is known as the "knee of the curve." In Figure 1.5 it is the point (0.5, 0.5). As Samson says (I agree with him):

> Unfortunately, most functions of interest resemble the M/M/1 queueing function shown in Figure 3 [our Figure 1.4].
>
> With a function like M/M/1, there is no critical zone in the domain of the independent variable. The choice of a guideline number is not easy, but the rule-of-thumb makers go right on.
>
> In most cases, there is not a knee, no matter how much we wish to find one. Rules of thumb must be questioned if offered without accompanying models that make clear the consequences of violation

Samson says "the germ of truth" about the second rule of thumb (Keep device utilization below 33%) is:

> If we refer to Figure 3, we see that when the M/M/1 model is an accurate representation of device queueing behavior, a device that is one-third busy will incur a queueing delay equal

> to half its service time. Someone decided many years ago that these numbers had some magical significance—that a device less than one-third busy wasn't busy enough, and that delay more than half of service time was excessive.

Samson has other wise things to say about this rule in his "The rest of the story" and "Lesson of the legend" comments. You may want to check that

$$\frac{1}{3} \times \frac{S}{(1-\frac{1}{3})} = \frac{S}{2}.$$

With respect to the third questionable rule of thumb (Keep path utilization below 30%), Samson points out that it is pretty much the preceding rule repeated. With newer systems, path utilizations exceeding 30% often have satisfactory performance. You must study the specific system rather than rely on questionable rules of thumb.

The final questionable rule of thumb (Keep CPU utilization below ??%) is the most common. The ?? value is usually 70 or 80. This rule of thumb overlooks the fact that it is sometimes very desirable for a computer system to run with 100% CPU utilization. An example is an interactive system that runs these workloads at a high priority but also has low priority batch jobs to utilize the CPU power not needed for interactive work. Rosenberg's three-part rule of thumb applies here.

1.2.4.2 Back-of-the-Envelope Modeling

Back-of-the-envelope modeling refers to informal calculations such as those that might be done on the back of an envelope if you were away from your desk. (I find *Mathematica* is very helpful for these kinds of calculations, if I am at my desk.) This type of modeling is often done as a rough check on the feasibility of some course of action such as adding 100 users to an existing interactive system. Such calculations can often reveal that the action is in one of three categories: Feasible with no problems, completely unfeasible, or a close call requiring more detailed study.

Petroski in his beautiful paper [Petroski 1991] on engineering design says:

> Back-of-the-envelope calculations are meant to reveal the reasonableness or ridiculousness of a design before it gets too far beyond the first sketch. For example, one can draw on the back of a cigarette box a design for a single-span suspension bridge

> between England and France, but a quick calculation on the same box will show that the cables, if they were to be made of any reasonable material, would have to be so heavy that they could not even hold up their own weight, let alone that of the bridge deck. One could also show that, even if a strong enough material for the cable could be made, the towers would have to be so tall that they would be unsightly and very expensive to build. Some calculations can be made so easily that engineers do not even need a pencil and paper. That is why the designs that they discredit are seldom even sketched in earnest, and serious designs proposed over the centuries for crossing the English Channel were either tunnels or bridges of many spans.

Similar remarks concerning the use of back-of-the-envelope calculations apply to the study of computer systems, of course. We use back-of-the-envelope calculations frequently throughout this book. For more about back-of-the-envelope modeling for computer systems see my paper [Allen 1987].

Exercise 1.1

Two women on bicycles face each other at opposite ends of a road that is 40 miles long. Ms. West at the western end of the road and Ms. East at the eastern end start toward each other, simultaneously. Each of them proceeds at exactly 20 miles per hour until they meet. Just as the two women begin their journeys a bumblebee flies from Ms. West's left shoulder and proceeds at a constant 50 miles per hour to Ms. East's left shoulder then back to Ms. West, then back to Ms. East, etc., until the two women meet. How far does the bumblebee fly? Hint: For the first flight segment we have the equation $50 \times t = 40 - 20 \times t$ where t is the time in hours for the flight segment. This equation yields t = 40/70 or a distance of 200/7 = 28.571428571 miles.

1.2.4.3 Statistical Projection

Many forms of statistical projection or forecasting exist. All of them use collected performance information from log files to establish a trend. This trend can then be projected into the future to predict performance data at a future time. Since some performance measures, such as response time, tend to be nonlinear it is difficult to use linear statistical forecasting to predict these measures except for short time periods. However, other statistical forecasting methods, such as exponential or S-curve, can sometimes be used. Other performance measures, such as utilization of a resource, tend to be nearly linear and thus can be projected more accurately by linear statistical methods.

Table 1.2. Mathematica Program

We enter the data.	In[4]:=cpu={0.605,0.597,0.623,0.632,0.647,0.639,0.676,0.723,0.698,0.743,0.759,0.772}
We plot the data.	In[6]:= gp=ListPlot[cpu]
Command for least squares fit.	In[8]:= g=N[Fit[cpu,{1,x},x],5] Out[8]= 0.56867 + 0.016538*x
Plot the fitted line.	In[9]:= Plot[g,{x,1,12}];
Plot points and line. See Figure 1.6.	In[10]:= Show[%,gp]

Linear Projection

Linear projection is a very natural technique to apply since most of us tend to think linearly. We believe we'd be twice as happy is we had twice as much money, etc. Suppose we have averaged the CPU utilization for each of the last 12 months to obtain the following 12 numbers {0.605, 0.597, 0.623, 0.632, 0.647, 0.639, 0.676, 0.723, 0.698, 0.743, 0.759, 0.772}. Then we could use the *Mathematica* program shown in Table 1.2 to fit a least-squares line through the points ; see Figure 1.6 for the result.

The least-squares line is the line fitted to the points so that the sum of the squares of the vertical deviations between the line and the given points is minimized. This is a straightforward calculation with some nice mathematical properties. In addition, it leads to a line that intuitively "looks like a good fit." The concept of a least-squares estimator was discovered by the great German mathematician Karl Friedrick Gauss in 1795 when he was 18 years old!

Figure 1.6. Linear Projection

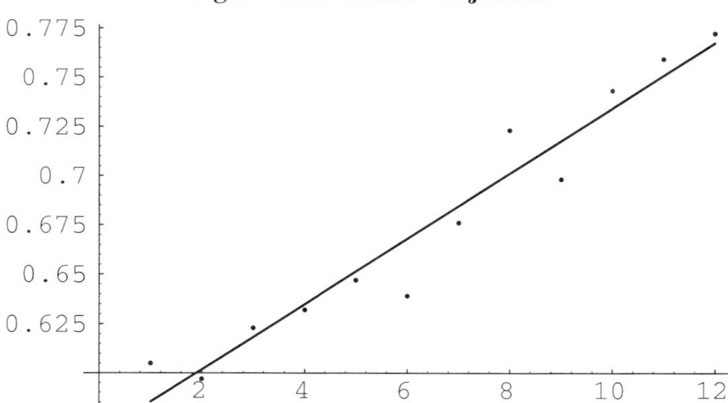

One must use great care when using linear projection because data that appears linear over a period of time sometimes become very nonlinear in a short time. There is a standard mathematical way of fitting a straight line to a set of points called *linear regression* which provides both (a) a measure of how well a straight line fits the measured points and (b) how much error to expect if we extend the straight line forward to predict values for the future. We will discuss these topics and others in the chapter on forecasting.

HP RXForecast Example

Figure 1.7 is an example of how linear regression and forecasting can be done with the Hewlett-Packard product HP RXForecast/UX. The figure is from page 2-16 of the HP RXForecast User's Manual for HP-UX Systems. The fluctuating curve is the smoothed curve of observed weekly peak disk utilization for a computer using the UNIX operating system. The center line is the trend line which extends beyond the observed values. The upper and lower lines provide the 90% prediction interval in which the predicted values will fall 90 percent of the time.

Other Statistical Projection Techniques

There are nonlinear statistical forecasting techniques that can be used, as well as the linear projection technique called linear regression. We will discuss these techniques in the chapter on forecasting.

Another technique is to use statistical forecasting to estimate future workload requirements. The workload estimates can then be used to parameterize a queueing theory model or a simulation model to predict the performance parameters such as average response time, average throughput, etc.

Figure 1.7. HP RXForecast/UX Example

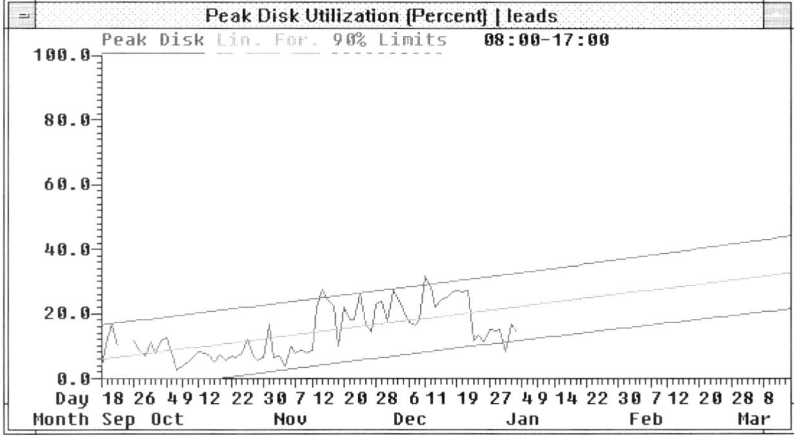

Business unit forecasting can be used to make computer performance estimates from business unit estimates. The business units used for this purpose are often called *natural forecasting units*, abbreviated as NFUs. Examples of NFUs are number of checking accounts at a bank, number of orders for a particular product, number of mail messages processed, etc. Business unit forecasting is a two step process. The first step is to use historical data on the business units and historical performance data to obtain the approximate relationship between the two types of data. For example, business unit forecasting might show that the number of orders received per day has a linear relationship with the CPU utilization of the computer system that processes the orders. In this case the relationship between the two might be approximated by the equation $U = 0.04 + 0.06 \times O$ where U is the CPU utilization and O is the number of orders received (in units of one thousand). Thus, if 12,000 orders were received in one day, the approximate CPU utilization is estimated to be 0.76 or 76%.

The second step is to estimate the size of the business unit at a future date and, from the approximate relationship, predict the value of the performance measure. In our example, if we predicted that the number of orders per day six months from today would be 15,000, then the forecasted CPU utilization would be $0.04 + 0.06 \times 15 = 0.94$ or 94%. We discuss this kind of forecasting in more detail in the chapter on forecasting.

Those with Hewlett-Packard computer systems can use HP RXForecast to perform all the statistical forecasting techniques we have discussed. We give examples of its use in the forecasting chapter.

1.2.4.4 Simulation Modeling

Bratley, Fox, and Schrage [Bratley, Fox, and Schrage 1987] define simulation as follows:

> Simulation means driving a model of a system with suitable inputs and observing the corresponding outputs.

Thus simulation modeling is a process that is much like measurement of an actual system. It is essentially an experimental procedure. In simulation we mimic or emulate an actual system by running a computer program (the simulation model) that behaves much like the system being modeled. We predict the behavior of the actual system by measurements made while running the simulation model. The simulation model generates customers (workload requests) and routes them through the model in the same way that a real workload moves through a computer system. Thus visits are made to a CPU representation, an I/O device representation, etc. The following basic steps are used:

1. Construct the model by choosing the service centers, the service center service time distributions, and the interconnection of the center.
2. Generate the transactions (customers) and route them through the model to represent the system.
3. Keep track of how long each transaction spends at each service center. The service time distribution is used to generate these times.
4. Construct the performance statistics from the preceding counts.
5. Analyze the statistics.
6. Validate the model.

Example 1.1

In this example we show that simulation can be used for other interesting problems that we encounter every day. The problem we discuss is called the "Monty Hall problem" on computer bulletin boards. Marilyn vos Savant, in her syndicated column "Ask Marilyn" published in the September 9, 1990, issue of *Parade,* asked the following question: "Suppose you're on a game show and you're given a choice of three doors. Behind one door is a car; behind the others, goats. You pick a door—say, No. 1—and the host, who knows what's behind the doors, opens another door—say, No. 3—which has a goat. He then says to you, 'Do you want to pick door No. 2?' Is it to your advantage to switch your choice?" Marilyn answered, "Yes, you should switch. The first door has a 1/3 chance of winning, but the second door has a 2/3 chance." Ms. vos Savant went on to explain why you should switch. It should be pointed out that the way the game host operates is as follows: If you originally pick the door with the car behind it, the host randomly picks one of the other doors, shows you the goat, and offers to let you switch. If you originally picked a door with a goat behind it, the host opens a door with a goat behind it and offers to let you switch. There was incredible negative response to the column leading Ms. vos Savant to write several more columns about the problem. In addition several newspaper articles and several articles in mathematical newsletters and journals have appeared. In her February 17, 1991, column she said:

> Gasp! If this controversy continues, even the *postman* won't be able to fit into the mailroom. I'm receiving thousands of letters, nearly all insisting that I'm wrong, including one from the deputy director of the Center for Defense Information and another from a research mathematical statistician from the National Institutes of Health! Of the letters from the general public, 92% are against my answer and of the letters from universities, 65% are against my answer. Overall, nine out of 10 readers completely disagree with my reply.

She then provides a completely convincing demonstration that her answer is correct and to suggest that children in schools set up a physical simulation of the problem. In her July 7, 199,1 column Ms. vos Savant published testimonials from grade school math teachers and students around the country who participated in an experiment that proved her right. Ms. vos Savant's columns are also printed in her book [vos Savant 1992]. We wrote the *Mathematica* simulation program **trial** which will simulate the playing of the game both with a player who never switches and another who always switches. Note that the first player wins only when his or her first guess is correct while the second wins whenever the first guess is incorrect. Since the latter condition is true two-thirds of the time, the switch player should win two-thirds of the time as Marilyn predicts. Let's let the program decide! The program and the output from a run of 10,000 trials are shown in Table 1.3.

Table 1.3. Mathematica Program

Name of program and parameter n.	`trial[n_] :=`
Initialize variables.	`Block[{switch=0, noswitch=0},`
Randomly choose n values of correct door.	`correctdoor=Table[Random[Integer, {1,3}], {n}];`
Randomly choose n values of first guess.	`firstchoice=Table[Random[Integer, {1,3}], {n}];`
Iterator.	`For[i=1, i<=n, i++,`
If switcher wins add to switcher total; otherwise add to noswitcher total.	`If[Abs[correctdoor[[i]]-firstchoice[[i]]]>0, switch=switch+1, noswitch=noswitch+1]];`
Return provides the fraction of wins for the switcher and nonswitcher.	`Return[{N[switch/n,8],N[noswitch/n,8]}];` `]`
	`In[4]:= trial[1000]`
	`Out[4]= {0.667, 0.333}`

The best and shortest paper in a mathematics or statistics journal I have seen about Marilyn's problem is the paper by Gillman [Gillman 1992]. Gillman also discusses some other equivalent puzzles. In the paper [Barbeau 1993], Barbeau discusses the problem, gives the history of the problem with many references, and considers a number of equivalent problems.

We see from the output that, with 10,000 trials, the person who always switches won 66.7% of the time and someone who never switches won 33.3% of the time for this run of the simulation. This is good evidence that the switching strategy will win about two-thirds of the time. Marilyn is right!

Several aspects of this simulation result are common to simulation. In the first place, we do not get the exact answer of 2/3 for the probability that a contestant who always switches will win, although in this case it was very close to 2/3. If we ran the simulation again we would get a slightly different answer. You may want to try it yourself to see the variability.

Don't feel bad if you disagreed with Marilyn. Persi Diaconis, one of the best known experts on probability and statistics in the world—he won one of the famous MacArthur Prize Fellowship "genius" awards—said about the Monty Hall problem, "I can't remember what my first reaction to it was because I've known about it for so many years. I'm one of the many people who have written papers about it. But I do know that my first reaction has been wrong time after time on similar problems. Our brains are just not wired to do probability problems very well, so I'm not surprised there were mistakes."

Exercise 1.2

This exercise is for programmers only. If you do not like to write code you will only frustrate yourself with this problem.

Consider the land of Femina where females are held in such high regard that every man and wife wants to have a girl. Every couple follows exactly the same strategy: They continue to have children until the first female child is born. Then they have no further children. Thus the possible birth sequences are G, BG, BBG, BBBG,.... Write a *Mathematica* simulation program to determine the average number of children in a family in Femina. Assume that only single births occur—no twins or triplets, every family does have children, etc.

1.2.4.5 Queueing Theory Modeling

This modeling technique represents a computer system as a network of service centers, each of which is treated as a queueing system. That is, each service center has an associated queue or waiting line where customers who cannot be served immediately queue (wait) for service. The customers are, of course, part of the queueing network. Customer is a generic word used to describe workload requests such as CPU service, I/O service requests, requests for main memory, etc. A simulation model also thinks of a computer system as a network of queues. Simplifying assumptions are made for analytic queueing theory models so that a solvable system of equations can be used to approximate the system modeled. Analytical queueing theory modeling is so well developed that most computer systems can be successfully modeled by them. Simulation models are more general than analytical models but require a great deal more effort to set up, validate, and run. We will demonstrate the use of both kinds of models later in this book.

Modeling is used not only to determine when the current system needs to be upgraded but also to evaluate possible new configurations. Boyse and Warn [Boyse and Warn 1975] provided one of the first documentations of the successful use of analytic queueing theory models to evaluate the possible configuration changes to a computer system. The computer system they were modeling was a mainframe computer with a virtual memory operating system servicing automotive design engineers who were using graphics terminals. These terminals put a heavy computational load on the system and accessed a large database. The system supported 10 terminals and had a fixed multiprogramming level of three, that is, three jobs were kept in main memory at all times. The two main upgrade alternatives that were modeled were: (a) adding 0.5 megabytes of main memory (computer memory was very expensive at the time this study was made) or (b) procuring I/O devices that would reduce the average time required for an I/O operation from 38 milliseconds to 15.5 milliseconds. Boyse and Warn were able to show that the two alternatives would have almost the same effect upon performance. Each would reduce the average response time from 21 to 16.8 seconds, increase the throughput from 0.4 to 0.48 transactions per second, and increase the number of terminals that could be supported with the current average response time from 10 to 12.

1.2.4.6 Simulation Versus Analytical Queueing Theory Modeling

Simulation and analytical queueing theory modeling are competing methods of solving queueing theory models of computer systems.

Simulation has the advantage of allowing more detailed modeling than analytical queueing theory but the disadvantage of requiring more resources in terms of development effort and computer resources to run. Queueing theory models are easier to develop and use less computer resources but cannot solve some models that can be solved by simulation.

Calaway [Calaway 1991] compares the two methods for the same study. The purpose of the study was to determine the effect a proposed DB2 application [DB2 (Data Base 2) is a widely used IBM relational database system] on their computer installation. The study was first done using the analytic queueing theory modeling package Best/1 MVS from BGS Systems, Inc. and then repeated using the simulation system SNAP/SHOT that is run by IBM for its customers. The system studied was a complex one. As Calaway says:

> The configuration studied was an IBM 3090 600E that was physically partitioned into two IBM 3090 300Es. Each IBM 3090 300E was logically partitioned using PR/SM into two logical machines. Side A consisted of processor 2 and proces-

> sor 4. Side B consisted of processor 1 and processor 3. This article compares the results of SNAP/SHOT and BEST/1 based on the workload from processor 2 and processor 4. The workload on these CPUs included several CICS regions, batch, TSO, ADABAS, COMPLETE and several started tasks. The initial plan was to develop the DB2 application on the processor 4 and put it into production on processor 3.

Calaway's conclusion was:

> The point is that for this particular study, an analytical model was used to reach the same acquisition decision as determined by a simulator and in a much shorter time frame (3.5 days vs. seven weeks) and with much less effort expended. I have used BEST/1 for years to help make acquisition decisions and I have always been pleased with the outcome.

It should be noted that the simulation modeling would have taken a great deal longer if it had been done using a general purpose simulation modeling system such as GPSS or SIMSCRIPT. SNAP/SHOT is a special purpose simulator designed by IBM to model IBM hardware and to accept inputs from IBM performance data collectors.

1.2.4.7 Benchmarking

Dongarra, Martin, and Worlton [Dongarra, Martin, and Worlton 1987] define benchmarking as "Running a set of well-known programs on a machine to compare its performance with that of others." Thus it is a process used to evaluate the performance or potential performance of a computer system for some specified kind of workload. For example, personal computer magazines publish the test results obtained from running benchmarks designed to measure the performance of different computer systems for a particular application such as word processing, spread sheet analysis, or statistical analysis. They also publish results that measure the performance of one computer performing the same task, such as spread sheet analysis or statistical analysis, with different software systems; this type of test measures software performance rather than hardware performance. There are standard benchmarks such as Livermore Loops, Linpack, Whetstones, and Dhrystones. The first two benchmarks are used to test scalar and vector floating-point performance. The Whetstones benchmark tests the basic arithmetic performance of midsize and small computers while the Dhrystones benchmark tests the nonnumeric performance of midsize and smaller computers. Much better benchmark suites have been developed by three new organizations: the Standard Performance Evaluation Corporation (SPEC), the Transaction Processing Performance Council (TPC), and the Business Applications Performance Corporation (BAPCo). These organizations and their benchmarks are discussed in Chapter 6.

No standard benchmark is likely to represent accurately the workload of a particular computer installation. Only a benchmark built specifically to test the environment of the computer installation can do that. Unfortunately, constructing such a benchmark is very resource intensive, very time consuming, and requires some very special skills. Only companies with large computer installations can afford to construct their own benchmarks. Very few of these companies use benchmarking because other modeling methods, such as analytic queueing theory modeling, have been found to be more cost effective. For a more complete discussion see [Incorvia 1992].

We discuss benchmarking further in Chapter 6.

1.2.5 Validation

Before a model can be used for making performance predictions it must, of course, be validated. By validating a model we mean confirming that it reasonably represents the computer system it is designed to represent.

The usual method of validating a model is to use measured parameter values from the current computer system to set up and run the model and then to compare the predicted performance parameters from the model with the measured performance values. The model is considered valid if these values are close. How close they must be to consider the model validated depends upon the type of model used. Thus a very detailed simulation model would be expected to perform more accurately than an approximate queueing theory network model or a statistical forecasting model. For a complex simulation model the analyst may need to use a statistical testing procedure to make a judgment about the conformity of the model to the actual system. One of the most quoted papers about statistical approaches to validation of simulation models is [Schatzoff and Tillman 1975]. Rules of thumb are often used to determine the validity of an approximate queueing theory model. Back-of-the-envelope calculations are valuable for validating any model. In all validation procedures, common sense, knowledge about the installed computer system, and experience are important.

Validating models of systems that do not yet exist is much more challenging than validating a model of an existing system that can be measured and compared with a model. For such systems it is useful to apply several modeling techniques for comparison. Naturally, back-of-the-envelope calculations should be made to verify that the model output is not completely wrong. Simulation is the most likely modeling technique to use as the primary technique but it should be cross-checked with queueing theory models and even simple benchmarks. A talent for good validation is what separates the outstanding modelers from the also-rans.

1.2.6 The Ongoing Management Process

Computer installations managed under service level agreements (SLAs) must be managed for the long term. Even installations without SLAs should not treat computer performance management as a "one-shot" affair. To be successful, performance management must be a continuing effort with documentation of what happens over time not only with a performance database but in other ways as well. For example, it is important to document all assumptions made in performance predictions. It is also important to regularly compare predictions of the performance of an upgraded computer system to the actual observed performance of the system after the upgrade is in place. In this way we can improve our performance predictions—or find someone else to blame in case of failure.

Another important management activity is defining other management goals as well as performance goals even for managers who are operating under one or more SLAs. System managers who are not using SLAs may find that some of their goals are a little nebulous. Typical informal goals (some goals might be so informal that they exist only inside the system manager's head) might be:

1. Keep the users happy.
2. Keep the number of performance complaint calls below 10 per day.
3. Get all the batch jobs left at the end of the first shift done before the first shift the next morning.

All system managers should have the first goal—if there were no users there would be no need for system managers! The second goal has the virtue of being quantified so that its achievement can be verified. The last goal could probably qualify as what John Rockart [Rockart 1979] calls a *critical success factor*. A system manager who fails to achieve critical success factor goals will probably not remain a system manager for very long. (A critical success factor is something that is of critical importance for the success of the organization.)

Deese [Deese 1988] provides some interesting comments on the management perspective on capacity planning.

Exercise 1.3

You are the new systems manager of a departmental computer system for a marketing group at Alpha Alpha. The system consists of a medium-sized computer connected by a LAN to a number of workstations. Your customers are a number of professionals who use the workstations to perform their daily work. The previous systems manager, Manager Manager (he changed his name from John Smith to Manager Manager to celebrate his first management position), left things in a chaotic mess. The users complain about

1. Very poor response time—especially during peak periods of the day, that is, just after the office opens in the morning and in the middle of the afternoon.
2. Unpredictable response times. The response time for the same application may vary between 0.5 seconds and 25 seconds even outside the busiest periods of the day!
3. The batch jobs that are to be run in the evening often have not been processed when people arrive in the morning. These batch jobs must be completed before the marketing people can do their work.

   (a) What are your objectives in your new job?
   (b) What actions must you take to achieve your objectives?

Exercise 1.4

The following service level agreement appears in [Duncombe 1991]:

SERVICE LEVEL AGREEMENT

THIS AGREEMENT dated August 6, 1991 is entered into by and between

The Accounts Payable Department, a functional unit of Acme Screw Enterprises Inc. (hereinafter called 'AP')

```
WITNESSETH that in consideration of the mutual
covenants contained herein, the parties agree as
follows:

1.EXPECTATIONS
The party of the first part ('AP') agrees to limit their
demands on and use of the services to a reasonable
level.

The party of the second part ('MIS') agrees to provide
computer services at an acceptable level.

2. PENALTIES
If either party to this contract breaches the
aforementioned EXPECTATIONS, the breaching party must
buy lunch.

IN WITNESS WHEREOF the parties have executed this
agreement as of the day and year first above written.

By:      ______________________________
Title:  ______________________________
Witness:______________________________
Date:    ______________________________


What are the weaknesses of this service level agreement?
How could you remedy them?
```

1.2.7 Performance Management Tools

Just as a carpenter cannot work without the tools of the trade–hammers, saws, levels, etc.–computer performance analysts cannot perform without proper tools. Fortunately, many computer performance management tools exist. The most common tool is the *software monitor*, which runs on your computer system to collect system resource consumption data and reports performance metrics such as response times and throughput rates.

There are four basic types of computer performance tools which match the four aspects of performance management shown in Figure 1.1.

Diagnostic Tools

Diagnostic tools are used to find out what is happening on your computer system now. For example, you may ask, "Why has my response time deteriorated from 2 seconds to 2 minutes?" Diagnostic tools can answer your question by telling you what programs are running and how they are using the system resources. Diagnostic tools can be used to discover problems such as a program caught in a loop and burning up most of the CPU time on the system, a shortage of memory causing memory management problems, excessive file opening and closing causing unnecessary demands on the I/O system, or unbalanced disk utilization. Some diagnostic monitors can log data for later examination.

The diagnostic tool we use the most at the Hewlett-Packard Performance Technology Center is the HP GlancePlus family. Figure 1.8 is from the HP GlancePlus/UX User's Manual [HP 1990]. It shows the last of nine HP GlancePlus/UX screens used by a performance analyst who was investigating a performance problem in a diskless workstation cluster.

Figure 1.8. HP GlancePlus/UX Example

```
HP B1791A  GlancePlus/UX A.00.00  10:55:11 client13 9000/340 Current  Avg  High
--------------------------------------------------------------------------------
CPU Util    S           SUU                                 | 34%   25%   99%
Disk Util                                                   |  0%    0%    0%
Memory Util S              SU                              U| 98%   98%  100%
Swap Util   R                                  R            | 80%   80%   96%
--------------------------------------------------------------------------------
                              MEMORY DETAIL
Event              Count  Cumulative  Current Rate   Avg Rate   High Rate
--------------------------------------------------------------------------------
Page Faults          660       14121          20.4       15.1        34.6
Page In/Outs         638       12003          19.7       13.1        31.7
KB Paged In         4788       93529         147.8      100.1       229.2
KB Paged Out        1160       17098          35.8       21.1       108.8
Swap In/Outs           4        1415           0.1        0.1         1.0
KB Swapped In         48        1340           1.5        1.4        14.4
KB Swapped Out        84        4711           2.6        4.9        47.6
VM Reads               0           0           0.0        0.0         0.0
VM Writes              0           0           0.0        0.0         0.0
Cache Hits             1         565        100.0%      84.7%      100.0%

Phys Memory:     4mb     Avail Memory:   2.4mb       Free Memory:   0.0mb

Global    CPU     Memory    Disk          Next    Select      Help      Exit
                                          Keys    Process               Glance
```

By "diskless workstation cluster" we mean a collection of workstations on a LAN that do not have local hard disk drives; a file server on the LAN takes care of the I/O needs of the workstations. One of the diskless workstation users had reported that his workstation was performing very poorly. Figure 1.8 indicates that the paging and swapping levels are very high. This means there is a severe memory bottleneck on the workstation. The "Physical Memory" line on the screen shows that the workstation has only 4 MB of memory. The owner of this workstation is a new user on the cluster and does not realize how much memory is needed.

Resource Management Tools

The principal resource management tool is a software monitor that monitors and logs system resource consumption data continuously to provide an archive or database of historical performance data. Companion tools are needed to manipulate and analyze this data. For example, as we previously mentioned, the software monitor provided by Hewlett-Packard for all its computer systems is the SCOPE monitor, which collects and summarizes performance data before logging it. HP LaserRX is the tool used to retrieve and display the data using Microsoft Windows displays. Other vendors who market resource management tools for Hewlett-Packard systems are listed in the Institute for Computer Management publication [Howard].

For IBM mainframe installations, RMF is the most widely used resource management tool. IBM provides RMF for its mainframes supporting the MVS, MVS/XA, and MVS/ESA operating systems. RMF gathers and reports data via three monitors (Monitor I, Monitor II, and Monitor III). Monitor I and Monitor II measure and report the use of resources. Monitor I is used mainly for archiving performance information while Monitor II primarily measures the contention for systems resources and the delay of jobs that such contention causes. Monitor III is used mostly as a diagnostic tool. Some of the third parties who provide resource management tools for IBM mainframes are Candle Corporation, Boole & Babbage, Legent, and Computer Associates. Most of these companies have overall system monitors as well as specialized monitors for heavily used IBM software such as CICS (Customer Information Control System), IMS (Information Management System), and DB2 (Data Base 2). For detailed information about performance tools for all manufacturers see the Institute for Computer Management publication [Howard].

Application Optimization Tools

Program profilers, which we discussed earlier, are important for improving code efficiency. They can be used both proactively, during the software development process, or reactively, when software is found to consume excessive amounts of computer resources. When used reactively program profilers (sometimes called program analyzers) are used to isolate the performance problem areas in the code. Profilers can be used to trace program execution, provide the statistics on system calls, provide information on computer resources consumed per transaction (CPU time, disk I/O time, etc.), time spent waiting on locks, etc. With this information the application can be tuned to perform more efficiently. Unfortunately, program profilers and other application optimization tools seem to be the Rodney Dangerfields of software tools; they just don't get the respect they deserve. Software engineers tend to feel that they know how to make a program efficient without any outside help. (Donald Knuth, regarded by many, including myself, to be the best programmer in the world, is a strong believer in profilers. His paper [Knuth 1971] is highly regarded by knowledgable programmers.) Literature is limited on application optimization tools, and even computer performance books tend to overlook them. An exception is the excellent introduction to profilers provided by Bentley in his chapter on this subject [Bentley 1988]. Bentley provides other articles on improving program performance in [Bentley 1986].

The neglect of profilers and other application optimization tools is unfortunate because profilers are available for most computers and most applications. For example, on an IBM personal computer or plug compatible, Borland International, Inc., provides Turbo Profiler, which will profile programs written using Turbo Pascal, any of Borland's C++ compilers, and Turbo Assembler, as well as programs compiled with Microsoft C and MASM. Other vendors also provide profilers, of course. Profilers are available on most computer systems. The profiler most actively used at the Hewlett-Packard Performance Technology Center is the HP Software Performance Tuner/XL (HP SPT/XL) for Hewlett-Packard HP 3000 computers. This tool was developed at the Performance Technology Center and is very effective in improving the running time of application programs. One staff member was able to make a large simulation program run in one-fifth of the original time after using HP SPT/XL to tune it. HP SPT/XL has also been used very effectively by the software engineers who develop new versions of the HP MPE/iX operating system.

Figure 1.9 displays a figure from page 3-4 of the HP SPT/XL User's Manual: Analysis Software. It shows that, for the application studied, 94.4% of the processing time was spent in system code. It also shows that DBGETs, which are calls to the TurboImage database system, take up 45.1% of the processing time. As can be seen from the DBGETS line, these 6,857 calls spend only a fraction of this time utilizing the CPU; the remainder of the time is spent waiting for something such as disk I/O, database locks, etc. Therefore, the strategy for optimizing this application would require you to determine why the application is waiting and to fix the problem.

Application optimization tools are most effective when they are used during application development. Thus these tools are important for SPE (systems performance engineering) activities.

Figure 1.9. HP SPT/XL Example

```
HPB1777 A.00.00 SPT/XL                                   Logfile Summary Information

Logfile: SPTLOG1.PUB.SPTDEMO
                                 Time(sec)     % of Total Processing Time

        Application code      12.597    5.6% |*                                       |
             System code     214.011   94.4% |*                                     * |
        Total processing time  226.609

        Intrinsic Name    Calls   Time(sec)     % of Total Processing Time

                 DBGET     6857   102.301   45.1% |C*           *                     |
            VREADFIELDS    1261    32.212   14.2% |CC*                                |
               FCONTROL     260    21.153    9.3% |**                                 |
             VSHOWFORM     1271    11.698    5.2% |C                                  |
          VPUTIEEEREAL      500     7.708    3.4% |                                   |
          HPSWITCHTOCM      250     7.639    3.4% |                                   |
            VFIELDEDITS     755     7.254    3.2% |                                   |
                 DBPUT      250     5.238    2.3% |                                   |
              VPUTDINT      750     4.134    1.8% |                                   |
            VCLOSETERM        1     1.940    0.9% |                                   |

Would you like to define transactions(YES)?>
```

Capacity Planning Tools

Many of the tools that are used for resource management are also useful for capacity planning. For example, it is essential to have monitors that continuously record performance information and a database of performance information to do capacity planning. Tools are also needed to predict future workloads (forecasting tools). In addition, modeling tools are needed to predict the future performance of the current system as the workload changes as well as to predict the performance of the predicted workload with alternative configurations. The starting point of every capacity planning project is a well-tuned system so application optimization tools are required as well.

All the tools used for capacity planning are also needed for (SPE.

Expert Systems for Computer Performance Analysis

As Deese says in his insightful paper [Deese 1990]:

> An expert system is a computer program that emulates the way that people solve problems. Like a human expert, an expert system give advice by using its own store of knowledge that

> relates to a particular area of expertise. In expert systems terminology, the knowledge generally is contained in a **knowledge base** and the area of expertise is referred to as a **knowledge domain**. The expert system's knowledge often is composed of both (1) facts (or conditions under which facts are applicable) and (2) heuristics (i.e., "rules of thumb").
>
> With most expert systems, the knowledge is stored in "IF/THEN" rules that describe the circumstances under which knowledge is applicable. These expert systems usually have increasingly complex rules or groups of rules that describe the conditions under which diagnostics or conclusions can be reached. Such systems are referred to as "rule-based" expert systems.
>
> Expert systems are used today in a wide variety of fields. These uses range from medical diagnosis (e.g., MYCIN[1]) to geological exploration (e.g., PROSPECTOR[2]), to speech EARSAY-II[3]), to laboratory instruction (e.g., SOPHIE[4]). In 1987, Wolfgram, et al, listed over 200 categories of expert system applications, with examples of existing expert systems in each category. These same authors estimate that by 1995, the expert system field will be an industry of over $9.5 billion!

Finally, in the last several years, expert systems for computer performance evaluation have been developed. As Hood says [Hood 1992]: "The MVS operating system and its associated subsystems could be described as the most complex entity ever developed by man." For this reason a number of commercial expert systems for analyzing the performance of MVS have been developed including **CA-ISS/THREE**, **CPExpert**, **MINDOVER MVS**, and **MVS Advisor**. **CA-ISS/ THREE** is especially interesting because it is one of the earliest computer performance systems with an expert system component as well as queueing theory modeling capability.

In his paper [Domanski 1990] Domanski cites the following advantages of expert systems for computer performance evaluation:

1. Expert systems are often cost effective when human expertise is very costly, not available, or contradictory.

2. Expert systems are objective. They are not biased to any pre-determined goal state, and they will not jump to conclusions.

3. Expert systems can apply a systematic reasoning process requiring a very large knowledge base that a human expert cannot retain because of its size.
4. Expert systems can be used to solve problems when given an unstructured problem or when no clear procedure/algorithm exists.

Among the capabilities that have been implemented by computer performance evaluation expert systems for mainframe as well as smaller computer systems are problem detection, problem diagnosis, threshold analysis, bottleneck analysis, "what's different" analysis, prediction using analytic models, and equipment selection. "What's different" analysis is a problem isolation technique that functions by comparing the attributes of a problem system to the attributes of the same system when no problem is present. The differences between the two sets of measurements suggest the cause of the problem. This technique is discussed in [Berry and Heller 1990].

The expert system **CPExpert** from Computer Management Sciences, Inc., is one of the best known computer performance evaluation expert systems for IBM or compatible mainframe computers running the MVS operating system. **CPExpert** consists of five different components to analyze different aspects of system performance. The components are SRM (Systems Resource Manager), MVS, DASD (disk drives in IBM parlance are called DASD for "direct access storage devices), CICS (Customer Information Control System), and TSO (Time Sharing Option). We quote from the Product Overview:

> **CPExpert** runs as a normal batch job, and it:
> Reads information from your system to detect performance problems.
>
> Consolidates and analyzes data from your system (normally contained in a performance database such as MXG™ or MICS™®) to identify the causes of performance problems.
>
> Produces narrative reports to explain the results from its analysis and to suggest changes to improve performance.
>
> **CPExpert** is implemented in SAS®, and is composed of hundreds of expert system rules, analysis modules, and queueing models. SAS was selected as our "expert system shell" because of its tremendous flexibility in summarizing, consolidating, and analyzing data. **CPExpert** consists of over 50,000 SAS statements, and the number of SAS statements increases regu-

larly as new features are implemented, new options are provided, or additional analysis is performed.

CPExpert has different components to analyze different aspects of system performance.

The **SRM Component** analyzes SYS1.PARMLIB members to identify problems or potential problems with your IPS or OPT specifications, and to provide guidance to the other components. Additionally, the SRM Component can convert your existing Installation Performance Specifications to MVS/ESA SP4.2 (or SP4.3) specifications.

The **MVS Component** evaluates MVS in the major MVS controls (multiprogramming level controls, system paging controls, controls for preventable swaps, and logical swapping controls).

The **DASD Component** identifies DASD volumes with the most significant performance problems and suggests way to correct the problems.

The **CICS Component** analyzes CICS statistics, applying most of the analysis described in IBM's CICS Performance Guides.

The **TSO Component** identifies periods when TSO response is unacceptable, "decomposes" the response time, and suggests way to reduce TSO response.

From this discussion it is clear that an expert system for a complex operating system can do a great deal to help manage performance. However, even for simpler operating systems, an expert system for computer performance analysis can do a great deal to help manage performance. For example, Hewlett-Packard recently announced that an expert system capability has been added to the online diagnostic tool HP GlancePlus for MPE/iX systems. It uses a comprehensive set of rules developed by performance specialists to alert the user whenever a possible performance problem arises. It also provides an extensive online help facility developed by performance experts. We quote from the HP GlancePlus User's Manual (for MPE/iX Systems):

What Does The Expert Facility Do?

The data displayed on each GlancePlus screen is examined by the Expert facility, and any indicators that exceed the normal range for the size of system are highlighted. Since the highlighting feature adds a negligible overhead, it is permanently enabled.

A global system analysis is performed based on data obtained from a single sample. This can be a response to an on-demand request (you pressed the X key), or might occur automatically following each screen update, if the Expert facility is in continuous mode. During global analysis, all pertinent systemwide performance indicators are passed through a set of rules. These rules were developed by top performance specialists working on the HP 3000. The rules were further refined through use on a variety of systems of all sizes and configurations. The response to these rules establishes the degree of probability that any particular performance situation (called a symptom) could be true.

If the analysis is performed on demand, any symptom that has a high enough probability of being true is listed along with the reasons (rules) why it is probably the case, as in the following example:

```
XPERT Status: 75%  CHANCE  OF  GLOBAL  DISC  BOTTLE-
NECK.
Reason: PEAK UTIL    > 90.00 (96.4)
```

This says that "most experts would agree that the system is experiencing a problem when interactive users consume more than 90% of the CPU." Currently, interactive use is 96.4%. Since the probability is only 75% (not 100%), some additional situations are not true. (In this case, the number of processes currently starved for the CPU might not be high enough to declare a real emergency.)

...

High level analysis can be performed only if the Expert facility is enabled for high level—use the V command: XLEVEL=-HIGH. After the global analysis in which a problem type was not normal, the processes that executed during the last interval are examined. If an action can be suggested that might improve the situation, the action is listed as follows:

```
XPERT: Status 75% CHANCE OF GLOBAL CPU BOTTLENECK.
Reason: INTERACTIVE > 90.00 (96.4)
Action: QZAP pin 122 (PASXL) for MEL.EELKEMA from "C"
to "D" queue.
```

Action will not be instituted automatically since you may or may not agree with the suggestions.

The last "Action" line of the preceding display means that the priority should be changed (QZAP) for process identification number 122, a Pascal compilation (PASXL). Furthermore, the Log-on of the person involved is Mel.Eelkema, and his process should be moved from the C queue to the D queue. Mel is a software engineer at the Performance Technology Center. He said the expert system caught him compiling in an interactive queue where large compilations are not recommended.

The expert system provides three levels of analysis: low level, high level, and dump level. For example, the low level analysis might be:

```
XPERT Status: 50% CHANCE OF DISC BOTTLENECK.
Reason: PEAK UTIL   > 90.00 (100.0)
XPERT Status: 100% CHANCE OF SWITCH RATE PROBLEM.
Reason: SWITCH RATE   > HIGH LIMIT    (636.6)
```

If we ask for high level analysis of this problem, we obtain more details about the problems observed and a possible solution as follows:

```
XPERT Status: 50% CHANCE OF DISC BOTTLENECK.
Reason: PEAK UTIL   > 90.00 (100.0)
XPERT Status: 100% CHANCE OF SWITCH RATE PROBLEM.
Reason: SWITCH RATE > HIGH LIMIT  (636.6)
XPERT Dump Everything Level Detail:
------------------------------DISC Analysis----------
General DISC starvation exists in the C queue but no
unusual processes are detected. This situation is most
likely caused by the combined  effect of many pro-
cesses.
No processes did an excessive amount of DISC IO.
The following processes appear to be starved for DISC
IO:
```

```
You might consider changing the execution priority
or    rescheduling  processes to allow them to run.

JSNo Dev Logon          Pin Program Pri CPU%  Disc Trn
Resp Wait
S21  32 ANLYST.PROD     111 QUERY     C 17.9%  10.0   0
0.0  64%

------------------------------SWITCH Analysis---------
Excessive Mode Switching exists for processes in the D
queue.
An excessive amount of mode switching was found for
the following processes:
Check for possible conversion CM to NM or use the OCT
program
JSNo Dev Logon      Pin Program    Pri  CPU%  Disc  CM%
NMsw CMsw  J9   10 FIN.PROD   110 CHECKS     D   16.4%
2.3   0% 533    0
```

Processes (jobs) running under the control of the Hewlett-Packard MPE/iX operating system can run in compatibility mode (CM) or native mode (NM). Compatibility mode is much slower but is necessary for some processes that were compiled on the MPE/V operating system. The SWITCH analysis has discovered an excessive amount of mode switching and suggested a remedy.

The preceding display is an example of high level analysis. We do not show the dump level, which provides detail level on all areas analyzed by the expert system.

Expert systems for computer performance analysis are valuable for most computer systems from minicomputers to large mainframe systems and even supercomputers. They have a bright future.

1.3 Organizations and Journals for Performance Analysts

Several professional organizations are dedicated to helping computer performance analysts and managers of computer installations. In addition most computer manufacturers have a user's group that is involved with all aspects of the use the vendor's product, including performance. Some of the larger users groups have special interest subgroups; sometimes there is one specializing in performance. For example, the IBM Share and Guide organizations have performance committees.

The professional organization that should be of interest to most readers of this book is the Computer Measurement Group, abbreviated CMG. CMG holds a conference in December of each year. Papers are presented on all aspects of computer performance analysis and all the papers are available in a proceedings. CMG also publishes a quarterly, *CMG Transactions*, and has local CMG chapters that usually meet once per month. The address of CMG headquarters is The Computer Measurement Group, 414 Plaza Drive, Suite 209, Westmont, IL 60559, (708)655-1812-Voice, (708)655-1813-FAX.

The *Capacity Management Review*, formerly called *EDP Performance Review*, is a monthly newsletter on managing computer performance. Included are articles by practitioners, reports of conferences, and reports on new computer performance tools, classes, etc. It is published by the Institute for Computer Capacity Management, P. O. Box 82847, Phoenix, AZ 85071, (602)997-7374.

Another computer performance analysis organization that is organized to support more theoretically inclined professionals such as university professors and personnel from suppliers of performance software is *ACM Sigmetrics*. It is a special interest group of the *Association for Computing Machinery* (ACM). *Sigmetrics* publishes the *Performance Evaluation Review* quarterly and holds an annual meeting. One issue of the *Performance Evaluation Review* is the proceedings of that meeting. Their address is *ACM Sigmetrics*, c/o *Association of Computing Machinery*, 11 West 42nd Street, New York, NY 10036, (212) 869-7440.

1.4 Review Exercises

The review exercises are provided to help you review this chapter. If you aren't sure of the answer to any question you should review the appropriate section of this chapter.

1. Into what four categories is performance management segmented by the Hewlett-Packard Performance Technology Center?

2. What is a profiler and why would anyone want to use one?

3. What are the four parts of a successful capacity planning program?

4. What is a service level agreement?

5. What are some advantages of having a chargeback system in place at a computer installation? What are some of the problems of implementing such a system?

6. What is software performance engineering and what are some of the problems of implementing it?

7. What are the primary modeling techniques used for computer performance studies?

8. What are the three basic components of any computer system according to Rosenberg?

9. What are some rules of thumb of doubtful authenticity according to Samson?

10. Suppose you're on a game show and you're given a choice of three doors. Behind one door is a car; behind the others, goats. You pick a door—say, No. 1—and the host, who knows what's behind the doors, opens another door—say, No. 3—which has a goat. He then says to you, 'Do you want to pick door No. 2?' Is it to your advantage to switch your choice?

11. Name two expert systems for computer performance analysis.

1.5 Solutions

Solution to Exercise 1.1

This is sometimes called the von Neumann problem. John von Neumann (1903—1957) was the greatest mathematician of the twentieth century. Many of those who knew him said he was the smartest person who ever lived. Von Neumann loved to solve back-of-the-envelope problems in his head. The easy way to solve the problem (I'm sure this is the way you did it) is to reason that the bumblebee flies at a constant 50 miles per hour until the cyclists meet. Since they meet in one hour, the bee flies 50 miles. The story often told is that, when John von Neumann was presented with the problem he solved it almost instantly. The proposer then said, "So you saw the trick." He answered, "What trick? It was an easy infinite series to sum." Recently, Bailey [Bailey 1992] showed how von Neumann might have set up the infinite series for a simpler version of the problem. Even for the simpler version setting up the infinite series is not easy.

Solution to Exercise 1.2

We named the following program after Nancy Blachman who suggested a somewhat similar exercise in a *Mathematica* course I took from her and in her book [Blachman 1992] (I had not seen Ms. Blachman's solution when I wrote this program.).

```
nancy[n_]:=
Block[{i,trials, average,k},
(* trials counts the number of births *)
(* for each couple. It is initialized to zero. *)
trials=Table[0, {n}];
For[i=1, i<=n, i++,
While[True,trials[[i]]=trials[[i]]+1;If[Random[Integer
, {0,1}]>0,
Break[]] ];];
(* The while statement counts number of births *)
(* for couple i. *)
(* The while is set up to test after a pass through *)
(* the loop *)
(* so we can count the birth of the first girl baby. *)
average=Sum[trials[[k]], {k, 1, n}]/n;
Print["The average number of children is ", average];
]
```

It is not difficult to prove that, if one attempts to perform a task which has probability of success p each time one tries, then the average number of attempts until the first success is $1/p$. See the solution to Exercise 4, Chapter 3, of [Allen 1990]. Hence we would expect an average family size of 2 children. We see below that with 1,000 families the program estimated the average number of children to be 2.007—pretty close to 2!

```
In[8]:= nancy[1000]
                                        2007
The average number of children is       ----
                                        1000

In[9]:= N[%]

Out[9]= 2.007
```

This answer is very close to 2. Ms. Blachman sent me her solution before her book was published. I present it here with her problem statement and her permission. Ever the instructor, she pointed out relative to my solution: "By the way it is not necessary to include {0, 1} in the call to Random[Integer, {0, 1}]. Random[-Integer] returns either 0 or 1." The statement of her exercise and the solution from page 296 of [Blachman 1992] follow:

10.3 Suppose families have children until they have a boy. Run a simulation with 1000 families and determine how many children a family will have on average. On average, how many daughters and how many sons will there be in a family?

```
makeFamily[]:=
     Block[{
              children = { }
         },
         While[Random[Integer] == 0,
              AppendTo[children, "girl"]
       ];
          Append[children, "boy"]
   ]
makeFamily::usage="makeFamily[ ] returns a list of
children."
numChildren[n_Integer] :=
    Block[{
             allChildren
           },
           allChildren = Flatten[Table[makeFamily[ ],
{n}]];
           {
             avgChildren -> Length[allChildren]/n,
             avgBoys -> Count[allChildren, "boy"]/n,
             avgGirls -> Count[allChildren, "girl"]/n
           }
       ]
numChildren::usage="numchildren[n] returns statistics
on
        the number of children from n families."
```

You can see that Ms. Blachman's programs are very elegant indeed! It is very easy to follow the logic of her code. Her **numChildren** program also runs faster than my **nancy** program. I ran her program with the following result:

```
In[9]:= numChildren[1000]//Timing

Out[9]= {1.31*Second, {avgChildren -> 1019/500, avg-
Boys -> 1,
    avgGirls -> 519/500}}
```

I believe you will agree that 1019/500 is pretty close to 2.

The following program was written by Rick Bowers of the Hewlett-Packard Performance Technology Center. His program runs even faster than Nancy Blachman's but doesn't do quite as much.

```
girl[n_]:=
Block[ {boys=0},
For[i=1, i<=n, i++, While[Random[Integer]==0, boys=-
boys+1]];
Return[N[(boys+n)/n]]]
```

Solution to Exercise 1.3

The problems you face are, unfortunately, very common for managers of computer systems.

(a): We hope your objectives include one or both of the following:

1. Get the computer system functioning the way it should so that your users can be more productive.
2. Establish a symbiotic relationship with the users of your computer system, possibly leading to a service level agreement.

(b): Activities that are important to achieving these objectives include:

1. Finding the source of the difficulties with response time and the batch jobs not being run on time. This book is designed to help you solve problems like these.
2. Once the source of the problems is uncovered then the solutions can be undertaken. We hope this book will help with this, too.
3. You must communicate to your users what the reasons are for their poor service in the past and how you are going to fix the problems. It is important to keep the users apprised of what you are doing to remedy the problems and what the current performance is. The latter is usually in the form of a weekly or monthly performance report. The contents and format of the report will depend upon what measurement and reporting tools are available.

Solution to Exercise 1.4

The point of Duncombe's excellent article [Duncombe 1991] is that everything in the agreement must be specified unambiguously. As Duncombe says, these items include:

1. the parties involved
2. the definition of all the terms used in the agreement
3. the exact expectations of the parties
4. how the service level will be measured
5. how the service level will be monitored and reported
6. duration of the agreement
7. method of resolving disputes
8. how the contract will be terminated

For an excellent example of a service level agreement with notes on what the terms mean see [Dithmar, Hugo, and Knight 1989].

1.6 References

1. Arnold O. Allen, *Probability, Statistics, and Queueing Theory with Computer Science Applications,* Second Edition, Academic Press, San Diego, 1990.
2. Arnold O. Allen, "Back-of-the-envelope modeling," *EDP Performance Review*, July 1987, 1–6.
3. Rex Backman, "Performance contracts," *INTERACT*, September 1990, 50–52.
4. David H. Bailey, "A capacity planning primer," *SHARE 62 Proceedings*, 1984,
5. Herbert R. Bailey, "The girl and the fly: a von Neumann legend," *Mathematical Spectrum*, 24(4), 1992, 108—109.
6. Peter Bailey, "The ABCs of SPE: software performance engineering," *Capacity Management Review*, September 1991.
7. Ed Barbeau, "The problem of the car and goats," *The College Mathematics Journal,* 24(2), March 1993, 149—154.

8. Jon Bentley, *Programming Pearls*, Addison-Wesley, Reading, MA, 1986.

9. Jon Bentley, *More Programming Pearls*, Addison-Wesley, Reading, MA, 1988.

10. Robert Berry and Joseph Hellerstein, "Expert systems for capacity management," *CMG Transactions*, Summer 1990, 85—92.

11. Nancy Blachman, *Mathematica: A Practical Approach*, Prentice Hall, Englewood Cliffs, NJ, 1992.

12. John W. Boyse and David R. Warn, "A straightforward model for computer performance prediction," *ACM Computing Surveys*, June 1975, 73–93.

13. Paul Bratley, Bennett L. Fox, and Linus E. Schrage, *A Guide to Simulation,* Second Edition, Springer-Verlag, New York, 1987

14. Janet Butler, "Does chargeback show where the buck stops?," *Software*, April 1992, 48—59.

15. **CA-ISS/THREE**, Computer Associates International, Inc., Garden City, NY.

16. James D. Calaway, "SNAP/SHOT VS BEST/1," *Technical Support*, March 1991, 18—22.

17. Dave Claridge, "Capacity planning: a management perspective," *Capacity Management Review*, August 1992, 1—4.

18. *CMG, CMG Transactions*, Summer 1990. Special issue on expert systems for computer performance evaluation.

19. **CPExpert**, Computer Management Sciences, Inc., Alexandria, VA.

20. **DASD Advisor**, Boole & Babbage, Inc., Sunnyvale, CA.

21. Donald R. Deese, "Designing an expert system for computer performance evaluation, *CMG '88 Conference Proceedings*, Computer Measurement Group, 1988a, 75—80.

22. Donald R. Deese, "A management perspective on computer capacity planning," *EDP Performance Review*, April 1988b, 1—4.

23. Donald R. Deese, "An expert system for computer performance evaluation, *CMG Transactions*, Summer 1990, 69—75.

24. Hans Dithmar, Ian St. J. Hugo, and Alan J. Knight, *The Capacity Management Primer*, Computer Capacity Management Services Ltd., London, 1989.

25. Bernard Domanski, "An expert system's tutorial for computer performance evaluation," *CMG Transactions*, Summer 1990, 77—83.

26. Jack Dongarra, Joanne L. Martin, and Jack Worlton, "Computer benchmarking: paths and pitfalls," *IEEE Spectrum*, July 1987, 38–43.

27. Brian Duncombe, "Service level agreements-only as good as the data," INTEREX Proceedings, 1991, 5134-1—5134-12.

28. Brian Duncombe, "Managing your way to effective service level agreements," *Capacity Management Review*, December 1992.

29. Peter J. Freimayer, "Data center chargeback—a resource accounting methodology," *CMG'88 Conference Proceedings,* Computer Measurement Group, 1988, 771—775.

30. Leonard Gillman, "The car and the goat," *American Mathematics Monthly*, January 1992, 3—7.

31. Doug Grumann and Marie Weston, "Analyzing MPE XL performance: What is normal?", *INTERACT*, August 1990, 42--58.

32. John L. Hennessy and David A. Patterson, *Computer Architecture: A Quantitative Approach,* Morgan Kaufmann, San Mateo, CA, 1990.

33. Linda Hood, "The use of expert systems technology in MVS," Part 1, *Capacity Management Review*, July 1992, 6—9, Part 2, *Capacity Management Review,* August 1992, 5—8.

34. Alan Howard, "Tools, teamwork defuse politics of performance," *Software*, April 1992a, 62—78.

35. Alan Howard, "The politics of performance: selling SPE to application developers," *CMG '92 Conference Proceedings*, Computer Measurement Group 1992b, 978—982.

36. Phillip C. Howard, Editor, *IS Capacity Management Handbook Series, Volume 1, Capacity Planning,* Institute for Computer Capacity Management, updated every few months.

37. Phillip C. Howard, Editor, *IS Capacity Management Handbook Series, Volume 2, Performance Analysis and Tuning,* Institute for Computer Capacity Management, updated every few months.

38. *HP GlancePlus/UX User's Manual*, Hewlett-Packard, Mountain View, CA, 1990.

39. *HP GlancePlus User's Manual (for MPE/iX Systems)*, Hewlett-Packard, Roseville, CA, 1992.

40. Thomas F. Incorvia, "Benchmark cost, risks, and alternatives," *CMG '92 Conference Proceedings*, Computer Measurement Group, 1992, 895—905

41. Donald E. Knuth, "An empirical study of FORTRAN programs," *Software Practice and Experience*, 1(1), 1971, 105–133.

42. Doug McBride, "Service level agreements," *HP Professional,* August 1990, 58—67.

43. *Managing Customer Service,* Technical Report, Institute for Computer Capacity Management, 1989.

44. H. W. "Barry" Merrill, *Merrill's Expanded Guide to Computer Performance Evaluation Using the SAS System*, SAS, Cary, NC, 1984.

45. George W. (Bill) Miller, "Service Level Agreements: Good fences make good neighbors," CMG'87, Computer Measurement Group, 1987, 553—560.

46. **MINDOVER MVS**, Computer Associates International, Inc., Garden City, NY.

47. **MVS Advisor**, Domanski Sciences, Inc., 24 Shira Lane, Freehold, NJ, 07728.

48. Henry Petroski, "On the backs of envelopes," *American Scientist*, January-February 1991, 15–17.

49. John F. Rockart, "Chief executives define their own data needs," *Harvard Business Review*, March-April 1979, 81–93.

50. Jerry L. Rosenberg, "More magic and mayhem: formulas, equations, and relationships for I/O and storage subsystems," *CMG'91 Conference Proceedings,* Computer Measurement Group, 1991, 1136—1149.

51. Stephen L. Samson, "MVS performance management legends," *CMG '88 Conference Proceedings*, Computer Measurement Group, 1988, 148—159.

52. M. Schatzoff and C. C. Tillman, "Design of experiments in simulator validation," *IBM Journal of Research and Development*, 29(3), May 1975, 252–262.

53. William M. Schrier, "A comprehensive chargeback system for data communications networks," *CMG '92 Conference Proceedings*, Computer Measurement Group, 1992, 250—261.

54. Connie Smith, *Performance Engineering of Software Systems,* Addison-Wesley, Reading, MA, 1991.

55. Dennis Vanvick, "Getting to know U(sers): A quick quiz can reveal the depths of understanding—or misunderstanding—between users and IS," *ComputerWorld*, January 27, 1992, 103—107.

56. N. C. Vince, "Establishing a capacity planning facility," *Computer Performance*, 1(1), June 1980, 41–48.

57. Marilyn vos Savant, *Ask Marilyn*, St. Martin's Press, 1992.

58. Harry Zimmer, "Rules of Thumb '90," *CMG Transactions*, Spring 1990, 51–61.

Chapter 2 Components of Computer Performance

The cheapest, fastest, and most reliable components of a computer system are those that aren't there.
C. Gordon Bell

2.1 Introduction

In Chapter 1 we listed some of the hardware and software characteristics that had an effect on the performance of a computer system, that is, on how fast it will perform the work you want it to do. In this chapter we will consider these characteristics and some others in more detail. We also consider how these components or contributors to computer performance are modeled. In addition we shall attempt to give you a feeling for the relative size of the contributions of each of these components to the overall performance of a computer system in executing a workload.

Our first task it to describe how we state a speed comparison between two machines performing the same task. For example, when someone says "machine A is twice as fast as machine B in performing task X," exactly what is meant? We will use the definitions recommended by Hennessy and Patterson [Hennessy and Patterson 1990]. For example, "A is n% faster than machine B" means

$$\frac{\text{Execution Time}_B}{\text{Execution Time}_A} = 1 + \frac{n}{100},$$

where the numerator in the fraction is the time it takes machine B to execute task X and the denominator is the time it takes machine A to do so. Since we want to solve for n, we rewrite the formula in the form

$$n = \frac{\text{Execution Time}_B - \text{Execution Time}_A}{\text{Execution Time}_A} \times 100.$$

To avoid confusion we always set up the ratio so that n is positive, that is, we talk in terms of "A is faster than B" rather than "B is slower than A." Let us consider an example.

Example 2.1

A *Mathematica* calculation took 17.36 seconds on machine A and 74.15 seconds on machine B. Since $\frac{74.15}{17.36} = 4.2713 = 1 + \frac{327.13}{100}$ we say that machine A is 327.13% faster than machine B. The reader should check that the formula for n provided earlier gives the correct result.

An easier way to make the computation is to use the *Mathematica* program **perform**, which follows:

```
perform[A_, B_] :=
(* A is the execution time on machine A *)
(* B is the execution time on machine B *)
Block[{n, m},
 n = ((B-A)/A) 100;
 m = ((A-B)/B) 100;
If[A <= B,
Print["Machine A is n% faster than machine B where  n =
", N[n, 10]],
 Print["Machine B is n% faster than machine A where n =
", N[m, 10]]];
]
```

Applying **perform** to Example 2.1 yields:

```
In[6]:= perform[17.36, 74.15]
Machine A is n% faster than machine B where n =
327.1313364
```

It does not matter if you key in the input in the wrong order. Note that **perform** uses A to refer to the first input so that, if you key in the smaller number as the second input, **perform** will report that B is faster than A. As a review you might try the following exercise using **perform.**

Exercise 2.1

We know that machine A runs a program in 20 seconds while machine B requires 30 seconds to run the same program. Which of the following statements is true?

1. A is 50% faster than B.
2. A is 33% faster than B.
3. Neither of the above.

This completes the statement of the exercise.

Every discipline has some folklore attached to it; performance management follows this tradition. A story that is often heard is that of a computer installation that had execrable performance so the management team decided to get a more powerful central processing unit (CPU). Since the original performance bottleneck was the I/O system, which was not improved, the performance actually degraded because the new CPU could generate I/O requests faster than the old one!

What we want to look into now is the increase in speed that can be achieved by improving the performance of part of a computer system such as the CPU or the I/O devices. The key tool for this purpose is Amdahl's law. In their book [Hennessy and Patterson 1990], Hennessy and Patterson provide Amdahl's law in the form

$$\frac{\text{Execution Time}_{\text{old}}}{\text{Execution Time}_{\text{new}}} = \frac{1}{1 - \text{Fraction}_{\text{enhanced}} + \dfrac{\text{Fraction}_{\text{enhanced}}}{\text{Speedup}_{\text{enhanced}}}} = \text{Speedup}_{\text{overall}}.$$

This formula defines speedup and describes how we calculate it using Amdahl's law, the middle formula. Thus the speedup is two if the new execution time is exactly one half the old execution time. Let us consider an example.

Example 2.2

Suppose we are considering a floating-point coprocessor for our computer. Suppose, also, that the coprocessor will speed up numerical processing by a factor of 20 but that only 20% of our workload uses numerical processing. We want to compute the overall speedup from obtaining the floating-point coprocessor. We see that $\text{Fraction}_{\text{enhanced}} = 0.2$, and $\text{Speedup}_{\text{enhanced}} = 20$ so that

$$\text{Speedup}_{\text{overall}} = \frac{1}{0.8 + \frac{0.2}{20}} = 1.234568.$$

Amdahl's law is important in that it shows that, if an enhancement can only be used for a fraction of a job, then the maximum speedup cannot exceed the reciprocal of one minus that fraction. In Example 2.2, the maximum speedup is limited by the reciprocal of 0.8 or 1.25. This also demonstrates the law of diminishing returns; speeding up the coprocessor to 50 times as fast as the computer without it will improve the overall speedup very little over the 20 times speedup. (In fact, only from 1.2345679 to 1.2437811 or 0.75%.) The only thing that would really help the speedup would be to increase the fraction of the time that it is effective.

The *Mathematica* program **speedup** from the package **first.m** can be used to make speedup calculations. The listing of the program follows.

```
speedup[enhanced_, speedup_] :=
(* enhanced is percent of time in enhanced mode *)
(* speedup is speedup while in enhanced mode *)
Block[{frac, speed},
frac = enhanced / 100;
speed = 1 /(1 - frac + frac / speedup);
Print["The speedup is ", N[speed, 8]];
]
```

The *Mathematica* program **speedup** can be used to make the calculation in Example 2.2 as follows:

```
In[6]:= speedup[20, 20]
The speedup is 1.2345679
```

The speedup certainly has an interesting decimal expansion! If only there were an 8 before the 9. The computation for a coprocessor that will speed up numerical calculations by a factor of 50 follows:

```
In[4]:= speedup[20, 50]
The speedup is 1.2437811
```

The concepts of speedup and "A is n% faster than B" are related but not equivalent concepts. For example, if machine A is enhanced in such a way as to run 100% faster for all its calculations and called machine B, then the speedup as an enhanced system is 2.0 and machine B is 100% faster than machine A.

2.2 Central Processing Units

On most computer systems the CPU (CPUs on multiprocessor systems) is the basic determining factor for both the price of the system and the performance of the system in doing useful work. For example, when comparing the performance of a selection of PCs, say notebook computers, a PC journal, such as *PC Computing* or *PC Magazine,* will group them according to CPU power.

How do we measure CPU power? The short answer is, "With a great deal of difficulty." Let us consider the basic hardware first.

The CPU power is fundamentally determined by the *clock period*, also called *CPU cycle time* or *clock cycle*. It is the smallest unit of time in which the CPU can execute a single instruction. (According to [Kahaner and Wattenberg 1992] the Hitachi S-3800 has the shortest clock cycle of any commercial computer in the world; it is two billionths of a second!) On complex instruction set computer systems (CISC) such as PCs using Intel 80486 or Intel 80386 microprocessors, IBM mainframe computers, or any computer built more than 10 years ago, most instructions require multiple CPU cycles. By contrast, RISC (reduced instruction set computers) are designed so that most instructions execute in one CPU cycle. In fact, by using *pipelining*, most RISC machines can execute more than one instruction per clock cycle, on the average. Pipelining is a method of improving the throughput of a CPU by overlapping the execution of multiple instructions. It is described in detail in [Hennessy and Patterson 1990] and [Stone 1993]. It is described conceptually in [Denning 1993]. A machine that can issue multiple independent instructions per clock cycle (perform pipelining) is said to be *superscalar.* Basic CPU speed is specified by its clock rate, which is the number of clock cycles per second, but usually given in terms of millions of clock cycles per second or MHz. If the clock cycle time is 10 nanoseconds or $10 \times 10^{-9} = 10^{-8}$ seconds per cycle, then the clock rate is $10^8 = 100$ million cycles per second or 100 MHz. It is customary to use "ns" as an abbreviation for "nanosecond" or "nanoseconds." As these words are being written (June 1993), the fastest Intel 80486DX microprocessor available runs at 50 MHz. Intel has delivered two 486DX2 microprocessors. The 486DX2 microprocessor is functionally identical and completely compatible with the 486DX family. The DX2 chip adds something Intel calls speed-doubler technology—which means that it runs twice as fast internally as it does with components external to the chip. To date a 50 MHz chip and a 66 MHz chip are available. The 50 MHz version operates at internally while communicating externally with system components at 25 MHz. The 66 MHz version of the DX2 operates at 66 MHz internally and 33 MHz externally.

The Intel i586 microprocessor (code named the P5 until late October 1992 when Intel announced that it would be known as the Pentium) was released by Intel in March 1993. Personal computer vendors introduced and displayed personal computers using the Pentium chip in May 1993 at Comdex in Atlanta. As you are reading this passage you probably know all about the Pentium (i586) and possibly the i686 or i786. We can be sure that the computers available a year from any given time will be much more powerful than those available at the given time.

The clock rate can be used to compare two processors of exactly the same type, such as two Intel 80486 microprocessors, roughly but not exactly. Thus a 100 MHz Intel 80486 computer would run almost exactly twice as fast as a 50 MHz 80486, if the caches were the same size and speed, they each had the same amount of main memory of the same speed, etc. However, a computer with a 25 MHz Motorola 68040 microprocessor and the same amount of memory as a computer with a 25 MHz Intel 80486 microprocessor would not be expected to have the same computing power. The reason for this is that the average number of *clock cycles per instruction* (CPI) is not the same for the two microprocessors, and the CPI itself depends upon what program is run to compute it.

For a given program which has a given *instruction count* (number of instructions) or *instruction path length* (in the IBM mainframe world this is usually shortened to *path length*) the CPI is defined by the following equation $CPI = \frac{\text{CPU cycles for the program}}{\text{Instruction count for the program}}$. Thus the CPU time required to execute a program is given by the formula

$$\text{CPU time} = \text{Instruction count} \times \text{CPI} \times \text{Clock cycle time.}$$

In this formula, the instruction count depends upon the program itself, the instruction set architecture of the computer, and the compiler used to generate the instructions. Thus the CPI depends upon the program, the computer architecture, and compiler technology. The clock cycle time depends upon the computer architecture, that is, its organization and technology. Thus, not one of the three factors in the formula is independent from the other two! We note that the total CPU time depends very much upon what sort of work we are doing with our computer. Compiling a FORTRAN program, updating a database, and running a spreadsheet make very different demands upon the CPU.

At this point you are probably wondering, "Why has nothing been said about MIPS? Aren't MIPS a universal measure of CPU power?" In case you are not familiar with MIPS, it means "millions of instructions per second."

What is usually left out of the statement of the MIPS rating is what the instructions are accomplishing. Since computers require more clock cycles to perform some instructions than others, the number of instructions that can be executed in any time interval depends upon what mix of instructions is executed. Thus running different programs on the same computer can yield different MIPS. Thus there is no fixed MIPS rating for a given computer. Comparing different computers with different instruction sets is very difficult using MIPS because a program could require a great many more instructions on one machine than the other. One way that people have tried to get around this difficulty is to declare a certain computer as a standard and compare the time it takes to perform a certain task against the time it takes to perform it on the standard machine, thus generating relative MIPS. The machine most often used as a standard 1-MIPS machine is the VAX-11/780. (It is now widely known that the actual VAX-11/780 speed is approximately 0.5 MIPS.) For example, suppose program A ran on a standard-VAX-11/780 in 345 seconds but required only 69 seconds on machine B. Machine B would then be said to have a relative MIPS rating of 345/69 = 5. There are a number of obvious difficulties with this approach. If program A was written to run on an IBM 4381 or a Hewlett-Packard 3000 Series 955, it might be difficult to run the program on a VAX-11/780, so one would probably have to limit the use of this standard machine to comparisons with other VAX machines. Even then there would be the question of whether one should use the latest compiler and operating system on the VAX-11/780 or the original ones that were used when the rating was established. Weicker, the developer of the Dhrystone benchmark, in his paper [Weicker 1990], reported that he ran his Dhrystone benchmark program on two VAX-11/780 computers with different compilers. He reported that on the first run the benchmark was translated into 483 instructions that executed in 700 microseconds for a native MIPS rating of 0.69 MIPS. On the second run 226 instructions were executed in 543 microseconds, yielding 0.42 native MIPS. Weicker notes that the run with the lowest MIPS rating executed the benchmark faster.

In his paper Weicker addressed the question, "Why, then, should this article bother to characterize in detail these 'stone age' benchmarks?" (Weicker is referring to benchmarks such as the Dhrystone, Whetstone, and Linpack.) He answers in part:

> (2) Manufacturers sometimes base their MIPS rating on them. An example is IBM's (unfortunate) decision to base the published (VAX-relative) MIPS numbers for the IBM 6000 workstation on the old 1.1 version of Dhrystone. Subsequently, DEC and Motorola changed the MIPS computation rules for their

> competing products, also basing their MIPS numbers on Dhrystone 1.1.

What Weicker dislikes is that the Dhrystone 1.1 benchmark is run to obtain a rating in Dhrystones per second. This rating is then divided by 1757 to obtain the number of relative VAX MIPS. If you read that a computer manufacture claims a MIPS rating of, say, 50, with no further explanation, you can be almost certain that the rating was obtained in this way. Most manufacturers will also provide the results of the Dhrystone, Whetstone, and other leading benchmarks. As an example, I have a 33 MHz 80486DX personal computer. The Power Meter rating for my PC is 14.652 relative VAX MIPS. Power Meter (a product of The Database Group, Inc.) is a measurement program used by many PC vendors to obtain the relative VAX MIPS rating for their IBM PC or compatible computers.

Because of the difficulty in pinning down exactly what MIPS means, it is sometime said that, "MIPS means Meaningless Indication of Processor Speed."

The only meaningful measure of how fast your CPU can do your work is to use a monitor to measure how fast it does so. Of course your CPU also needs the assistance of other computer components such as I/O devices, cache, main memory, the operating system, etc., and no description of CPU performance is complete without specifying these other components as well. A typical software performance monitor will measure I/O activity as well as other indicators of information that is performance related.

Although there is some variability in how long it takes a CPU to perform even a simple operation, such as adding two numbers, there will be an averaging effect if you measure the performance of a computer system as it executes a program. The main problem is in selecting a program or mix of programs that faithfully represent the workload on your system. We discuss this problem in more detail in the chapter on benchmarking.

Example 2.3

Sam Spade has written a very clever piece of software called SeeItAll that will monitor the performance of any IBM PC or compatible computer. SeeItAll has magical properties; it provides any item of performance information that is of interest to anyone and causes no overhead on the PC measured. Using SeeItall, Sam measures the execution of the long *Mathematica* program **ComputeEverything** on his 50 MHz 80486 PC. He finds that **ComputeEverything** requires 50 seconds of CPU time and has an instruction count of 750 million instructions. What is the CPI for **ComputeEverything** on

Sam's machine? What is the MIPS rating of Sam's machine while running **ComputeEverything**?

Solution

The appropriate formula for the calculation is

$$\text{CPU time} = \text{Instruction count} \times \text{CPI} \times \text{Clock cycle time}.$$

To simplify the calculation we use *Mathematica* as follows:

```
In[3]:= Solve[750000000 CPI /(50 10^6) ==50]

Out[3]= {{CPI -> 10/3}}
```

This shows that the CPI is 10/3 clock cycles per instruction. Note that we used the formula

$$\text{Clock cycle time} = 1/\text{MHz} \times 10^6.$$

The MIPS rating is 750/50 or 15 because 750 million instructions were executed in 50 seconds. We can make these calculations easier using the *Mathematica* program **cpu** from the package **first.m**: .

```
cpu[instructions_, MHz_, cputime_] :=
(* instructions is number of instructions executed by
*)
(* the cpu in the length of time cputime *)
Block[{cpi,mips},
mips = 10^(-6) instructions / cputime;
cpi = MHz / mips;
Print["The speed in MIPS is ", N[mips, 8]];
Print["The number of clock cycles per instruction,
CPI, is ", N[cpi,10]];
]
```

Note that we use the identity $\text{CPI} = \text{MHz}/\text{MIPS}$. We left out the algebra that shows that this formula is true, but it follows from the formula

$$\text{CPU time} = \text{Instruction count} \times \text{CPI} \times \text{Clock cycle time}.$$

The calculations for Example 2.3 using **cpu** follow:

```
In[5]:= cpu[750000000, 50, 50]
The speed in MIPS is 15
The number of clock cycles per instruction, CPI, is
3.333333333
```

When using the formula CPI = MHz/MIPS it is very important to use an actual measured MIPS value and not the relative VAX MIPS value calculated from the results of a Dhrystone benchmark as described earlier.

Exercise 2.2

Sam Spade's friend Mike Hammer borrows SeeItAll to check the speed of the prototype of an IBM PC-compatible personal computer that his company is designing. He runs **ComputeEverything** in 20 seconds according to SeeItAll. Unfortunately, Mike doesn't know the speed of the Intel 80486 microprocessor in the machine. Could it be the 100 MHz microprocessor that everyone is talking about?

Exercise 2.3

Sam Spade's friend Dick Tracy claims that his company is designing an Intel 80486 clone with a clock speed of 200 MHz that will enable their new personal computer to execute the program **ComputeEverything** in 5 seconds flat. What CPI and MIPS are required for this machine to attain this goal?

The operation of a CPU with pipelineing, caching, and other advanced features is very difficult to model exactly. Fortunately, detailed modeling is not necessary for the purpose of performance management as it would be for engineers who are designing a new computer system. We need model only as accurately as we can predict future workloads. The CPU of a computer system can be effectively modeled with a queueing theory model using only the average amount of CPU service time required to run a representative workload. This number can be obtained from a software monitor. We discuss measurement considerations in Chapter 5.

So far we have discussed only uniprocessor systems, that is, computer systems with one CPU. Many computer systems have more than one processor and thus are known as *multiprocessor systems* (What else?). There are two basic organizations for such systems: loosely coupled and tightly coupled. Tightly coupled systems are more common. This type of organization is used for computer systems with a small number of processors, usually not more than 8, but 2 or 4 processors are more common. Loosely coupled systems usually have 32 or more processors. The new CM-5 Connection Machine recently announced by Thinking Machines has from 32 to 16,384 processors.

Tightly coupled multiprocessors, also called shared memory multiprocessors, are distinguished by the fact that all the processors share the same memory.

There is only one operating system, which synchronizes the operation of the processors as they make memory and database requests. Most such systems allow a certain degree of parallelism, that is, for some applications they allow more than one processor to be active simultaneously doing work for the same application. Tightly coupled multiprocessor computer systems can be modeled using queueing theory and information from a software monitor. This is a more difficult task than modeling uniprocessor systems because of the interference between processors. Modeling is achieved using a load dependent queueing model together with some special measurement techniques.

Loosely coupled multiprocessor systems, also known as *distributed memory systems*, are sometimes called *massively parallel computers* or *multicomputers.* Each processor has its own memory and sometimes a local operating system as well. There are several different organizations for loosely coupled systems but the problem all of them have is indicated by Amdahl's law, which says that the degree of speedup due to the parallel operation is given by

$$\text{Speedup} = \frac{1}{(1 - \text{Fraction}_{\text{parallel}}) + \dfrac{\text{Fraction}_{\text{parallel}}}{\text{n}}}$$

where n is the total number of processors. The problem is in achieving a high degree of parallelism. For example, if the system has 100 processors with all of them running in parallel half of the time, the speedup is only 1.9802. To obtain a speedup of 50 requires that the fraction of the time that all processors are operating in parallel is 98/99 = 0.98989899.

Thinking Machines is the best known company that builds massively parallel computers. Patterson, in his article [Patterson 1992], says of the latest Thinking Machines computer:

> In this historical context, the new Thinking Machines CM-5 may prove to be a landmark computer. The CM-5 bridges the two standard approaches to parallelism of the 1980s: single instruction, multiple data (SIMD) found in the CM-2 and MasPar machines, and multiple instruction, multiple data (MIMD) found in the Intel IPSC and Cray Y-MP.
>
> The single-instruction nature of SIMD simplifies the programming of massively parallel processors, but there are times when a single instruction stream is inefficient: when one of several operations must be performed based on the data, for

> example. An area where MIMD has the edge is in availability of components: MIMD machines can be constructed from the same processors found in workstations.
>
> The CM-5 merges these two styles by having two networks: one to route data, as found in all massively parallel machines, and another to handle the specific needs of SIMD (broadcasting information and global synchronization of processors). It also offers an optional vector accelerator for each processor. Hence the machine combines all three of the major trends in supercomputers: vector, SIMD, and MIMD.
>
> The CM-5 can be built around 32 to 16,384 nodes, each with an off-the-shelf RISC processor. Prices begin at about US$1 million and increase to well over $100 million for the largest version, which offers a claimed 1 teraflops in peak performance.
>
> Perhaps as important as the scaling of processor power, input/output (I/O) devices can also be easily integrated. Hence a CM-5 can be constructed with 1024 processors and 32 disks or 32 processors and 1024 disks, depending on the customer's needs.

Another very interesting massively parallel multiprocessor is the KSR-1 from Kendall Square Research in Cambridge, Massachusetts. The KSR-1 uses up to 1,088 64-bit microprocessors connected by a distributed memory scheme called ALLCACHE. This eliminates physical memory addressing so that work is not bound to a particular memory location but moves to the processors that require the data. The allure of the KSR-1 is that any processor can be deployed on either scalar or parallel applications. This makes it general purpose so that it can do both scientific and commercial processing. Gordon Bell, a computer seer, says [Bell 1992]:

> Kendall Square Research introduced their KSR 1 scalable, shared memory multiprocessors (smP) with 1,088 64-bit microprocessors. It provides a sequentially consistent memory and programming model, proving that smPs are feasible. The KSR breakthrough that permits scalability to allow it to become an ultracomputer is based on a distributed memory scheme, ALLCACHE, that eliminates physical memory addressing. The ALLCACHE design is a confluence of cache

> and virtual memory concepts that exploit locality required by scalable, distributed computing. Work is not bound to a particular memory, but moves dynamically to the processors requiring the data. A multiprocessor provides the greatest and most flexible ability for workload since any processor can be deployed on either scalar or parallel (e.g., vector) applications, and is general-purpose, being equally useful for scientific and commercial processing, including transaction processing, databases, real time, and command and control. The KSR machine is most likely the blueprint for future scalable, massively parallel computers.

This is truly an exciting time for computer designers and everyone who uses a computer will benefit!

There is a great deal of active research on parallel computing systems. The September/November 1991 issue of the *IBM Journal of Research and Development* is devoted entirely to parallel processing. Gordon Bell's paper [Bell 1992] is an excellent current technology review of the field. The papers [Flatt 1991], [Eager, Zahorjan, and Lazowska 1989], [Tanenbaum, Kaashoek, and Bal 1992], and [Kleinrock and Huang 1992] are excellent contemporary research papers on parallel processing. [Tanenbaum, Kaashoek, and Bal 1992] is an especially good paper for the software side of parallel computing. The September 1992 issue of *IEEE Spectrum* is a special issue devoted to supercomputers; it covers all aspects of the newest computer architectures as well as the problems of developing software to take advantage of the processing power. An update to some of the articles is provided in the January 1993 issue of *IEEE Spectrum*, the annual review of products and applications.

Ideally one would desire an indefinitely large memory capacity such that any particular word would be immediately available.... We are...forced to recognize the possibility of constructing a hierarchy of memories, each of which has greater capacity than the preceding but which is less quickly accessible.

A. W. Burks, H. G. Goldstine, and J. von Neumann

Preliminary Discussion of the Logical Design of an Electronic Computing Instrument(1946)

2.3 The Memory Hierarchy

Figure 2.1. The Memory Hierarchy

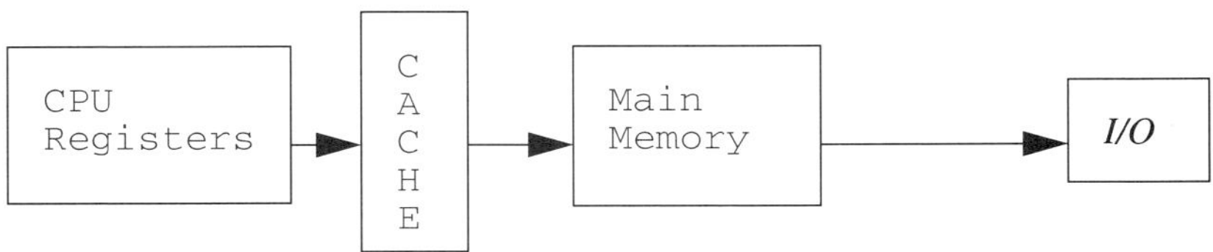

Figure 2.1 shows the typical memory hierarchy on a computer system; it is valid for most computers ranging from personal computers and workstations to supercomputers. It fits the description provided by Burks, Goldstine, and von Neumann in their prescient 1946 report. The fastest memory, and the smallest in the system, is provided by the CPU registers. As we proceed from left to right in the hierarchy, memories become larger, the access times increase, and the cost per byte decreases. The goal of a well-designed memory hierarchy is a system in which the average memory access times are only slightly slower than that of the fastest element, the CPU cache (the CPU registers are faster than the CPU cache but cannot be used for general storage), with an average cost per bit that is only slightly higher than that of the lowest cost element.

A CPU (processor) cache is a small, fast memory that holds the most recently accessed data and instructions from main memory. Some computer architectures, such as the Hewlett-Packard Precision Architecture, call for separate caches for data and instructions. When the item sought is not found in the cache, a *cache miss* occurs, and the item must be retrieved from main memory. This is a much slower access, and the processor may become idle while waiting for the data element to be delivered. Fortunately, because of the strong locality of reference exhibited by a program's instruction and data reference sequences, 95% to more than 98% of all requests are satisfied by the cache on a typical system. Caches work because of the *principle of locality.* The principle of locality is described by Hennessy and Patterson [Hennessy and Patterson 1990] as follows:

> This hypothesis, which holds that all programs favor a portion of their address space at any instant of time, has two dimensions:

> *Temporal locality* (locality in time)—If an item is referenced, it will tend to be referenced again soon.
>
> *Spatial locality* (locality in space)—If an item is referenced, nearby items will tend to be referenced soon.

Thus a cache operates as a system that moves recently accessed items and the items near them to a storage medium that is faster than main memory.

Just as all objects referenced by the CPU need not be in the CPU cache or caches, not all objects referenced in a program need be in main memory. Most computers (even Personal Computers) have virtual memory so that some lines of a program may be stored on a disk. The most common way that virtual memory is handled is to divide the address space into fixed-size blocks called pages. At any give time a page can be stored either in main memory or on a disk. When the CPU references an item within a page that is not in the CPU cache or in main memory, a *page fault* occurs, and the page is moved from the disk to main memory. Thus the CPU cache and main memory have the same relationship as main memory and disk memory. Disk storage devices, such as the IBM 3380 and 3390, have cache storage in the disk control unit so that a large percentage of the time a page or block of data can be read from the cache, obviating the need to perform a disk read. Special algorithms and hardware for writing to the cache have also been developed. According to Cohen, King, and Brady [Cohen, King, and Brady 1989] disk cache controllers can give up to an order of magnitude better I/O service time than an equivalent configuration of uncached disk storage.

Because caches consist of small high speed memory, they are very fast and can significantly improve the performance of computer systems. Let us see, in a rough sort of way, what a CPU cache can do for performance.

Example 2.4

Jack Smith has an older personal computer that does not have a CPU cache. He decides to upgrade his machine. The machine he decides is the best for him has two different CPU cache sizes available. Jack has used a profiler to study the large program that he uses most of the time. His calculations indicate that with the smallest of the two CPU caches he will get a cache hit 60% of the time while with the largest cache he will get a hit 90% of the time. How much will each of the caches speed up his processing compared to no cache at all if cache memory has a speedup of 5 compared to main memory?

Solution

We make the calculations with the *Mathematica* program **speedup** as follows:

```
In[9]:= speedup[60, 5]
The speedup is 1.9230769

In[10]:= speedup[90, 5]
The speedup is 3.5714286
```

Thus the smaller cache provides a speedup of 1.9230769 while the large cache speeds up the processing with speedup 3.5714286. It usually pays to obtain the largest cache offered because the difference in cost for a larger cache is usually small.

CPU caches make it more difficult to analyze benchmark results because many benchmark programs are so small that they fit into many caches although a typical program that is run on the system will not fit into the cache. Suppose, for example, your main application program had 20,000 lines of code and the 80/20 rule applied, that is, 20% of the code accounted for 80% of the execution time. Thus 4,000 lines of code account for 80% of the execution time. If the cache could hold 2,000 lines of code, then we would have a 40% hit rate for the CPU cache, that is, 50% of 80%. According to speedup, this would give us a speedup of 1.4705882:

```
In[8]:= speedup[40, 5]
The speedup is 1.4705882
```

The effect of the memory hierarchy on performance is the most difficult entity to model. Its main effect is to increase the variability of the time to process a transaction. This great variability is the result of the fact that the access to data on disk drives is a great deal slower than that of data in a CPU cache. For CPU caches memory access times are a few nanoseconds; the corresponding time to retrieve information from a disk drive is measured in milliseconds. We discuss disk drives in more detail in the section on input/output.

Main storage is a very important part of the memory hierarchy. In fact, most experienced computer performance analysts agree that "You cannot have too much main memory," and the corollary, "You can't have too much auxiliary memory, either." Joe Majors of IBM recommends: "Get the maximum main memory available; then increase slowly."

As Schardt says [Schardt 1980]:

> One characteristic of a system with a storage contention problem is the inability to fully utilize the processor. In some cases it may not be possible to get CPU utilization above 60 percent.
>
> The basic solution to a storage-constrained system is more real storage. If you have a four-megabyte IMS system and only three megabytes of storage to run it, no amount of parameter adjusting, System Resource Monitor modifications, or system zapping will make it run well. What will make it run well is four megabytes of storage, assuming the buffers have been tuned for system components such as TCAM, VTAM, VSAM, IMS, etc.

Some performance problems can only be cured by having enough memory. Fortunately, memory is becoming more inexpensive every year.

Let us consider an example of a system that you are probably familiar with that illustrates the memory hierarchy: my home personal computer, an IBM PC compatible with a 33 MHz Intel 80486DX microprocessor.

Example 2.5

The fastest memory in an IBM PC or compatible with a 33 MHz Intel 486DX microprocessor is in the CPU registers, which have access times of about 10 ns . The next fastest is the *primary cache* memory on the processor. Most 486 PCs also have an off chip cache called the *secondary cache*. Thus the primary cache is a cache into the secondary cache, which is a cache for main memory. This double caching is necessary because main memory speeds have not kept up with CPU speeds. Caches work because of the *principle of locality* described earlier. A cache operates as a system that moves recently accessed items and the items near them to a storage medium that is faster than main memory. The main memory access times for personal computers today (June 1993) varies from about 70 ns to 100 ns. The next level of storage below main memory is virtual storage, that is, hard disk storage. Hard disks typically have an access time of around 15 ms. This means that main memory is about 200,000 times as fast as hard disk memory. (On my PC this ratio is about 204,286.)

A significant problem with large, fast computers is that of providing sufficient I/O bandwidth to keep the CPU busy.

Richard E. Matick

IBM Systems Journal 1986

In an analysis of the components of response time, I/O time tends to be the dominant component, often accounting for 90 percent or more of the total.
Yogendra Singh, Gary M. King, and James W. Anderson, Jr.
IBM Systems Journal 1986

Because of its effect on the overall system throughput and end-user response time, minimization of DASD response time is a primary objective in the design of a storage hierarchy. ... Long-term trends in processor and DASD technology show a 10 percent compound increase of the processor and DASD-performance gap. Significant contributors to DASD performance are based on mechanical rather than electronic technologies. Therefore, other avenues must be explored to keep pace with the DASD response time requirements of systems.
Edward I. Cohen, Gary M. King, and James T. Brady
IBM Systems Journal 1989

2.3.1 Input/Output

I/O has been the Achilles' heel of computers and computing for a number of years, although there are some signs of improvement on the horizon. In fact Hennessy and Patterson, in their admirable book [Hennessy and Patterson 1990] have a chapter on Input/Output that begins with the paragraph:

> Input/output has been the orphan of computer architecture. Historically neglected by CPU enthusiasts, the prejudice against I/O is institutionalized in the most widely used performance measure, CPU time (page 35). Whether a computer has the best or the worst I/O system in the world cannot be measured by CPU time, which by definition ignores I/O. The second class citizenship of I/O is even apparent in the label "peripheral" applied to I/O devices.

They also say

> While this single chapter cannot fully vindicate I/O, it may at least atone for some of the sins of the past and restore some balance.

IBM refers to disk drives as *DASD* (for direct access storage devices) and disk memory is often referred to as *auxiliary storage* by most authors. PC users usually refer to their disk drives as *hard drives* or *fixed disks* to differentiate them from their floppy drives, which are used primarily to load new software or to back up the other drives.

Let us briefly review the characteristics of the most common I/O device on most computers from PCs and workstations to super computers: the magnetic disk drive. A magnetic disk drive has a collection of platters rotating on a spindle. The most common rotational speed is 3,600 revolutions per minute (RPM) although some of the newer drives spin at 6,400 RPM. The platters are metal disks covered with magnetic recording material on both sides. (Of course, the floppy drives on PCs have removable plastic disks called diskettes.) Disk drives have diameters as small as 1.8 inches for subnotebook computers and as large as 14 inches on mainframe drives such as the IBM 3990. (Hewlett-Packard announced a drive with a diameter of only 1.3 in in June 1992 with deliveries beginning in early 1993.)

The top as well as the bottom surface of each platter is used for storage and is divided into concentric circles called *tracks*. (On some drives, such as the IBM 3380, the top of the top platter and the bottom of the bottom platter are not used for storage.) A 1.44-MB floppy drive for a PC has 80 tracks on each surface; large drives can have as many as 2,200 tracks. Each track is divided into *sectors*; the sector is the smallest unit of information that can be read. A sector is 512 bytes on most disk drives. This is approximately the storage required for a half page of ordinary double-spaced text. A 1.44-MB floppy drive has 18 sectors per track; the 200-MB disk drive on my PC has 38 sectors on each of the 682 tracks on each of its 16 surfaces.

To read or write information into a sector, a read/write head is located over or under each surface attached to a movable *arm*. Bits are magnetically read or recorded on the track by the read/write head. The arms are connected so that each read/write head is over the same track of every device. A *cylinder* is the set of all tracks under the heads at a given time. Thus, if a disk drive has 20 surfaces, a cylinder consists of 20 tracks.

Each disk drive has a controller, which begins a read or write operation by moving the arm to the proper cylinder. This is called a *seek;* naturally the time required to move the read/write heads to the required cylinder is called the *seek time*. The minimum seek time is the time to move the arm one track; the maximum seek time is the time to move from the first to last track (or vice versa). The average seek time is defined by disk drive vendors as the sum of the time for all possible seeks divided by the number of possible seeks. However, due to locality of reference for most applications, in most cases measured average seek time is 25% to 30% of that provided by the vendors. (Sometimes no seek is required and large seeks are rarely required.) For example, Cohen, King, and Brady [Cohen, King, and Brady 1989] report: "The IBM 3380 Model K has a rated average seek time of 16 milliseconds. However, due to the reference pattern to the data, in most cases the experienced average seek is about 25 to 30 percent of the rated average seek."

Latency is the delay associated with the rotation of the platters until the requested sector is located under the read/write head. The average latency (usually called the latency) is the time it takes to complete a half revolution of the disk. Since most drives rotate at 3,600 RPM, the latency is usually 8.3 milliseconds.

The next component of the disk access time is the data transfer time. This is the time it takes to move the data from the storage device. It can be calculated by the formula

$$\frac{\text{number of sectors transferred}}{\text{number of sectors per track}} \times \text{disk rotation time} = \text{transfer time.}$$

For example, the 200-MB disk drive on my PC has 38 sectors, each 512 bytes long, for a total track capacity of 19,456 bytes. It rotates at 3,600 RPM and thus completes a rotation in 16.667 milliseconds or 0.016667 seconds. The time to transfer one sector of data is thus $1/38 \times 16.667 = 0.439$ milliseconds. The data transfer time is usually a small part of the access time. As Johnson says [Johnson 1991]: "For a 4,096–byte block on a 3.0 megabyte per second channel, it takes approximately 1.3 milliseconds for data transfer, yet performance tuning experts are happy when an average I/O takes 20 to 40 ms."

As we indicate in Figure 2.2, a string of disk drives is usually connected to the CPU through a channel and a control unit. Some IBM systems also have multiple strings connected to control units; each separate string of drives is connected through a head-of-string device.

Figure 2.2. An I/O System

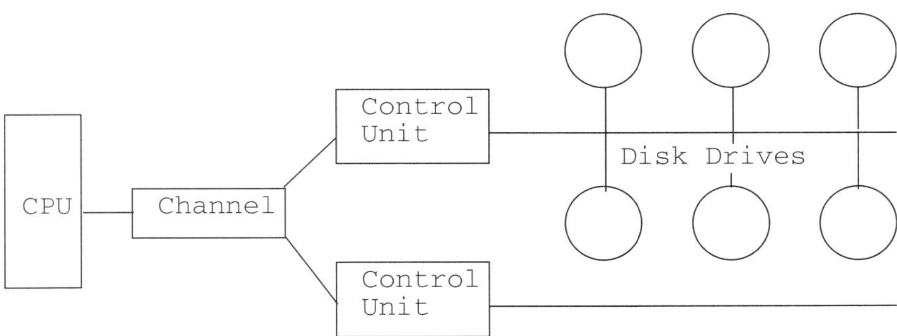

Rotational position sensing (RPS) is used for many I/O subsystems. This technique allows the transfer path (controller, channel, etc.) to be used by other devices during a drive's seek and rotational latency period. The controller tells the drive to issue an alert when the desired sector is approaching the read/write head. When the drive issues this signal, the controller attempts to establish a communication path to main memory so that the required data transfer can occur. If communication is established, the transfer is performed, and the drive is available for further service. If the attempt at connect fails because one or more of the path elements is busy, the drive must make a full revolution before another attempt at connection can be made. This additional delay is called an *RPS miss*. There are some drives such as the EMC Symmetrix II system which have actuator level buffers that eliminate RPS delay entirely. If a path is not available at the critical time, the information from the track is read into an actuator buffer. The information is then transmitted from the buffer when a path is available. This has the effect of lowering the channel utilization as well.

Some computer systems have alternative channel paths between the disk drives and the CPU. That is, each disk drive can be connected to more than one controller, and each controller can be connected to more than one channel. For these systems an RPS miss occurs only if all the channel paths are busy when the disk drive is ready to transmit data. On IBM systems this is called *dynamic path selection* (DPS) and up to four internal data paths are available for each disk drive. The DPS facility is sometimes known as "floating channels" because it allows a read command to a disk drive to go out on one channel while the data may be returned on a different channel.

The total disk access time is the sum of the seek time, latency time, transfer time, controller overhead, RPS miss time, and the queueing time. The queueing time is the most difficult to estimate and is the sum of the two delays: the initial delay until the drive is free so that it can be used and the delay until a channel is free to transmit the I/O commands to the disk. For non-RPS systems there is another queueing delay for the channel after the seek to place the read/write heads over the desired cylinder is completed. The channel is required to search for the sector to be read as well as for the transfer.

Example 2.6

Suppose Superdrive Inc. has announced a super new disk drive with the following characteristics: Average seek time 20 ms, rotation time 12.5 ms (4,800 RPM), and 150 sectors, each 512 bytes long, per track. Compute the average time to read or write an 8 sector block of data assuming no queueing delays, controller overhead of 2 ms, and no RPS misses.

Solution

The value of 2 ms for controller overhead is a value often used by I/O experts. Since we have assumed no queueing delays or RPS misses, the average time to access 8 sectors (4,096 bytes) is the sum of the average seek time, the average latency (rotational delay), data transfer time, and the controller overhead. We can safely use 30% of the average seek time provided by Superdrive or 6 ms for the average seek time. The average latency is 6.25 ms. By the formula we used earlier, the data transfer time is $(8/150) \times 12.5 = 0.6667$ ms. Hence the average access time is $6 + 6.25 + 0.6667 + 2 = 14.9167$ ms.

Exercise 2.4

Consider the following *Mathematica* program. Use **simpledisk** to verify the solution to Example 2.6.

```
simpledisk[seek_, rpm_, dsectors_, tsectors_, control-
ler_] :=
Block[{latency, transfer},
(* seek time in milliseconds, dsectors is number of
sectors per *)
(* track, tsectors is number of sectors to be trans-
ferred *)
(* controller is estimated controller time *)
Block[{latency, transfer, access},
latency = 30000/rpm;
```

```
transfer = 2 latency tsectors / dsectors;
access = latency + transfer + seek + controller;
Print["The latency time in milliseconds is ", N[la-
tency, 5]];
Print["The transfer time in milliseconds is ",
N[transfer, 6]];
Print["The access time in milliseconds is ", N[access,
6]];
]]
```

While I/O performance has not increased as much per year in recent years as CPU performance there have been some substantial improvements in disk performance, even on PCs. (Hennessy and Patterson claim it is 4% to 6% per year compared to 18% to 35% per year improvements in CPU performance.) Three years ago the average seek time for a PC hard disk was 28 ms or so. The hard disk I bought for my PC in May 1993 has an average seek time of 13.8 ms. The storage on this drive cost $1.39 per MB compared to $33.50 per MB for the RAM memory I bought at the same time. (These prices were about half what I spent for similar hardware in late 1991. They are probably even lower as you are reading this.)

Software and even hardware caching is often used on PCs, which further improves I/O performance. Even with these improvements I/O is still often the bottleneck.

This morning as I came into my office building I noticed a number of Hewlett-Packard HP7935 disk drives in the hall that were being replaced. (They look like the icon in the right margin.) These drives were state-of-the-art for HP 3000 computer systems in 1983 and only five years ago most computer rooms at Hewlett-Packard installations were full of them. (Some still are.) This drive which can store 404 MB of data is, according to my tape measure, 22 inches wide, 33 inches deep, and 32 inches high. The drives are usually stacked two high to produce a stack that is about the size of a phone booth. The average seek time on these drives is 24.0 ms with an average rotational delay of 11.1 ms. The drives I saw were replaced by Hewlett-Packard C2202A drives, which are stored in cabinets with four to each cabinet. These drives are the natural replacement for the HP7935s because they both use the HPIB interface. Hewlett-Packard has higher performance drives, which use the SCSI interface. Each C3302A drive can store 670 MB of data, has an average seek time of 17 ms and an average latency of 7.5 ms. Thus a cabinet that is much smaller than a HP7935 drive (14.5 in by 27 in by 28 in) can store 2.617 GB of data. The C2202A is a tremendous improvement over the HP7935 disk drive but not nearly as much improvement as there has been in CPU and memories over the period between the two drives. In

January 1993 Hewlett-Packard announced a drive that is 3 1/2 inches in diameter, stores 2.1 GB of data, has an access time of 8.9 ms and a spin rate of 6,400 RPM. Thus the latency is only 4.69 ms.

Larger computers have an even greater tendency than PCs to be reined in by the performance of the I/O subsystem. For example, IBM mainframes running the MVS operating system at one time had a reputation for poor I/O performance. In fact Lipsky and Church reported in their interesting modeling paper [Lipsky and Church 1977]:

> These studies indicate that the IBM 3330 disks are so much faster than the IBM 2314s that they can radically change the productivity of an IBM 360 computer—in fact, a good part of the superior productivity claimed for the IBM 370 may be due to the faster disks. Using faster disks on an IBM 360 can reduce the 20% to 30% idle time common for this machine to less than 10%.

In 1980 Schardt, an IBM engineer, reported [Schardt 1980] that:

> I/O contention, which in many cases is independent of the operating system in use, accounts for about 75 percent of the problems reported to the Washington Systems Center as poor MVS performance. Channel loading, control unit or device contention, data set placement, paging configurations, and shared DASD are often the major culprits.

In spite of these revelations IBM has never had anything but a good reputation for I/O design. Hennessy and Patterson say:

> If computer architects were polled to select the leading company in I/O design, IBM would win hands down. A good deal of IBM's mainframe business is commercial applications, known to be I/O intensive. While there are graphic devices and networks that can be connected to an IBM mainframe, IBM's reputation comes from disk performance.

Naturally, after these reports, IBM continued to improve its I/O performance. IBM increased the speed and size of its disk drives, added cache memory to the control units of some drives, and instituted "floating channels" so that the commands to read data from a disk drive could go out on one channel but the data retrieved could be returned on a different channel; hardware determines what channels to use. One of the biggest improvements was the announcement of the IBM 3090 with *expanded storage* which is also referred to in some IBM documents as *extended storage*. Expanded storage on the IBM 3090 and later models is not at all like expanded or extended storage on a personal computer; it is more like a RAM disk on a PC. Expanded storage on an IBM mainframe is generally regarded as an ultra-high-speed paging subsystem. When the MVS memory manager (called RSM for real storage manager although the IBM term for main memory is central storage) decides to move a page from main memory it can go either to disk storage (auxiliary memory) or to expanded storage. Similarly, when a page must be brought into main memory it can come from auxiliary storage or from expanded storage.

Expanded storage can only be used for 4K block transfers to and from central storage. Individual bytes in expanded storage cannot be addressed directly, and direct I/O transfers between expanded storage and conventional auxiliary storage cannot occur. The time to resolve a page fault for a page located in expanded storage can range from 75 to 135 microseconds (no one seems to be sure about the exact values of these ranges). This compares with an expected time of 2 to 20 milliseconds to resolve a page fault from auxiliary storage; thus expanded storage is from about 15 to 265 times as fast as auxiliary storage. There is also a savings in processor overhead for I/O initiation and the subsequent handling of the I/O completion interrupt.

There now seems to be a general perception that MVS I/O problems can be solved if adequate main and expanded storage is provided. As Beretvas says [Beretvas 1987]:

> Paging, as the key problem is rapidly disappearing for installations with adequate processor storage configurations. This is particularly true for IBM 3090 installations with expanded storage.

Samson [Samson 1992] claims that the MVS I/O problem has been solved for old applications but there are some new large applications now feasible because of the increased capabilities of the new IBM mainframes and the new releases of MVS\ESA; these new applications can create I/O performance problems.

In his paper [Artis 1992], Artis explains the evolution of the IBM I/O subsystem as it has evolved from the initial facilities provided by the IBM System/360 through the IBM System/390 system operating under MVS/ESA. An even more detailed discussion is presented in Chapter 1 of [Houtekamer and Artis 1992]. Artis has the following to say about the S/390 architecture:

> Introduced in September 1990, the primary objective of S/390 architecture relative to I/O was to address restrictions encountered during the end of the life of S/370-XA architecture. In particular, S/390 architecture introduced a new channel architecture called Enterprise Systems Connection (ESCON).
>
> ESCON architecture is based on 10MB and 17MB per second fiber optic channel technology that addresses both cable length and bandwidth restrictions that hampered large installations. In addition, the MVS/ESA operating system was updated to provide facilities for editing the IOCP of an active system. This capability addresses many of the nondisruptive installation requirements previously identified by MVS users.
>
> ...
>
> S/390 retains the distributed philosophy to I/O management introduced by S/370-XA architecture where EXDC was responsible for path selection and management of I/Os. Moreover, introduction of ESCON architecture and more powerful cached controllers will continue the trend to I/O decentralization.

Naturally, other computer manufacturers have similar stories to tell about the evolution of their I/O systems.

As we mentioned earlier, Hewlett-Packard has constantly improved their disk drives. For example, during 1991 the average seek time was reduced to 12.6 ms for the fastest drives. Most drives now have a latency of 7.5 ms or less and controller overhead has been lowered to less than 1 ms. In November 1991, Hewlett-Packard announced the availability of disk arrays, better known as RAID for Redundant Arrays of Inexpensive Disks (see [Patterson, Gibson, and Katz 1988]). (We discuss RAID later in this chapter.) In June 1992 Hewlett-Packard announced a disk drive with 21.4 MB of storage and a disk diameter of 1.3 in., thus becoming the first company to announce such a small disk drive. This amazing disk drive, called the Kittyhawk Personal Storage Module, is designed to withstand a system drop of about 3 feet during read/write operation. It spins at 5,400 RPM thus having a latency of 5.56 ms. It has an average seek time of less than 18 ms, a sustained transfer rate of 0.9 MB/second with a burst data rate of 1.2 MB/second. It has a spinup of approximately 1 second. One model (the one with 14 MB of storage) has one platter and two heads while the model with 21.4 MB of storage has two platters and three heads. This drive measures 0.4 in by 2 in by 1.44 in and weighs approximately 1ounce. Delivery of these drives began in early 1993. In March 1993 Hewlett-Packard announced a second version, the Kittyhawk II PSM, with a storage capacity of 42.8 MB. It remains the world's smallest disk drive and can store the equivalent of 28,778 typed pages of information.

In spite of the progress it has made with disk drives, Hewlett-Packard has recognized that the CPU and memory speeds on their computers are improving more rapidly than disk access speeds and that memory costs are constantly moving down. Therefore, Hewlett-Packard has improved the performance of I/O-intensive applications by increasing memory size and using main memory as a buffer for disk memory.

The HP 3000 MPE/iX operating system uses an improved disk caching capability called *mapped files*. The mapped files technique significantly improves I/O performance by reducing the number of physical I/Os without imposing additional CPU overhead or sacrificing data integrity and protection. This technique also eliminates file system buffering and optimizes global memory management.

Mapped files are based on the operating system's demand-paged virtual memory and are made possible by the extremely large virtual address space (MPE/iX provides approximately 281 trillion bytes of virtual address space) on the system. When a file is opened it is logically "mapped" into the virtual space. That is, all files on the system and their contents are referenced by virtual addresses. Every byte of each opened file has a unique virtual address.

File access performance is improved when the code and data required for processing can be found in main memory. Traditional disk caching reduces costly disk reads by using main memory for code and data. HP mapped files and virtual memory management further improve performance by caching writes. Once a virtual page is read into memory, it can be read by multiple users without additional I/O overhead. If it is a data page (HP pages data and instructions separately), it can be read and written to in memory without physically writing it to disk. When the desired page is already in memory, locking delays are greatly reduced, which increases throughput. Finally, when the memory manager does write a page back to disk, it combines multiple pages into a single write, again reducing multiple physical I/Os. The virtual-to-physical address translations to locate portions of the mapped-in files are performed by the system hardware, so that operating system overhead is greatly reduced.

In addition, the mapped file technique eliminates file system buffering. Since the memory manager fetches data directly into the user's area, the need for file system buffering is eliminated.

Other computer manufacturers have of course found other ways to improve their I/O performance. Companies that specialize in disk drives have been stretching the envelope over the last several years. In 1990, the typical, almost universal rotational speed of disk drives was 3,600 RPM. This has been increased to 4,004 RPM, then to 5,400 RPM, and, as we mentioned earlier, in January 1993 Hewlett-Packard announced a drive which a 6,400 RPM spin rate; thus its latency is only 4.69 ms. It also has 2.1 GB of storage capacity and a diameter of 3 1/2 in. You may be asking, "Why don't the mainframe folks speed up their large drives, too?" (Some mainframe drives have a diameter of 14 in.) The answer lies in physics. It is very difficult to keep a large drive from flying apart when it is spun rapidly. The smaller a drive, the faster it can spin. This is leading to small drives with very high data densities. By the time you read this paragraph the statistics of disk drive performance will surely be higher, but the improvements in disk technology will still be lagging the improvements in CPU and main memory speeds.

The hottest new innovation in disk storage technology is the *disk array*, more commonly denoted by the acronym *RAID* (Redundant Array of Inexpensive Disks). The seminal paper for this technology is the paper [Patterson, Gibson, and Katz 1988]. It introduced RAID terminology and established a research agenda for a group of researchers at the University of California at Berkeley for several years. The abstract of their paper, which provides a concise statement about the technology follows:

> Increasing performance of CPU and memories will be squandered if not matched by a similar performance increase in I/O. While the capacity of Single Large Expensive Disks (SLED) has grown rapidly, the performance improvement of SLED has been modest. Redundant Arrays of Inexpensive Disks (RAID), based on the magnetic disk technology developed for personal computers, offers an attractive alternative to SLED, promising improvements of an order of magnitude in performance, reliability, power consumption, and scalability. This paper introduces five levels of RAID, giving their relative cost/performance, and compares RAID to an IBM 3380 and a Fujitsu Super Eagle.

Lindholm in [Lindholm 1993] provides an excellent nontechnical introduction to the RAID technology including a sidebar called "Which RAID Is Right for Your App." This sidebar describes each RAID level and gives its pros and cons. Lindholm's paper also describes vendor extensions to the RAID technology to improve performance. An example is the Write Assist Drive (WAD) provided by IBM on the IBM 9337 to overcome RAID 5's write penalty. Lindholm also provides a selected list of RAID drive arrays available when the paper was published. Many of the key papers on RAID, including [Patterson, Gibson, and Katz 1988]are reprinted in [Friedman 1991]. As of August 1992 RAID in the form of the EMC Symmetrix 4416, 4424, and 4832 disk drives has been available on IBM mainframes running the MVS operating system for about a year. The devices appear to the system as an IBM 3380 or 3990 installation although it is faster and takes up much less floor space. According to an article in the June 15, 1992, issue of *ComputerWorld*, based on interviews with four companies using the devices, EMC's Symmetric models give users 50% faster response time than IBM's 3380 and 5% to 10% more speed than IBM's 3390. They require about one-fifth the floor space of conventional drives and cost about the same. EMC claims that Symmetrix I/O response times average 4 to 8 ms and throughputs of 1,500 to 2,000 I/Os per second can be achieved.

RAID storage products are traditionally compared to SLED (single, large, expensive disk) devices. RAID devices are faster, more reliable, and smaller than SLED devices. The speed is obtained by using very large caches and by reading or writing to a number of the disks in parallel. This parallel activity is called *striping*. It can be used because information is stored on a number of drives simultaneously. Striping provides a speed that is proportional to the number of drives used on one controller.

RAID reliability is obtained by using extra disks that contain redundant information that can be used to recover the original information when a disk fails. When a disk fails, it is assumed that within a short time the failed disk will be replaced and the information will be reconstructed on the new disk. There are six common levels of reliability available for RAID systems, running from level zero with simple striping to level five, which is a striping scheme with error correction codes. These levels are described in the classic paper [Patterson, Gibson, and Katz 1988]. The two most popular levels are Level 1 and Level 5. Level 1 provides mirrored disks. This is the most expensive option since all disks are duplicated and every write to a data disk is also a write to a check disk. It requires twice the storage space of a non-RAID solution compared to an average 20% overhead of RAID Level 5. It is also the fastest and most reliable level. Patterson, Gibson, and Katz in Table II of [Patterson, Gibson, and Katz 1988] show that with Level 1 and 10 to 25 disks it is possible to have a mean time to failure (MTTF) of over 500 years! The single most popular RAID organization is Level 5. Level 5 RAID distributes the data and check information across all the disks—including the check disks. As Patterson *et al.* say in [Patterson, Gibson, and Katz 1988]:

> These changes bring RAID level 5 near the best of both worlds: Small read-modify-writes now perform close to the speed per disk of a level 1 RAID while keeping the large transfer performance per disk and high useful storage capacity percentage of the RAID levels 3 and 4. Spreading the data across all disks even improves the performance of small reads, since there is one more disk per group that contains data. Table VI summarizes the characteristics of this RAID.

The paper [Patterson, Gibson, and Katz 1988] is an excellent introduction to RAID.

The three buzzwords that describe the methods of dealing with disk failure are *hot spares, hot fixes*, and *hot plugs*. A hot spare is an extra disk drive that is installed and running on the system but doing nothing until it is electronically switched on to take the place of a failed drive. The electronic switchover is called a *hot fix* and means that a failed disk drive can be replaced without shutting down the system. The hot plug technique means that the failed disk can be removed and replaced without shutting down the system.

By the time you read this passage RAID systems for PCs may be a reality! They actually are a reality now for PCs used as LAN file servers on as Bachus *et al.* describe [Bachus, Houston, and Longsworth 1993]. They tested seven RAID systems ranging in price from $12,500 to $37,995 for systems with between 2.2 GB and 8.0 GB of storage. Still a little above budget for most PCs not used as file servers. Prices are dropping rapidly. Quinlan [Quinlan 1993] reports that Hewlett-Packard has announced a disk array that is priced from $8,849 for a three-disk system with a RAID level 5 storage capacity of 1 GB to $14,899 for a five-disk array with a level 5 storage capacity of 4 GB. Perhaps I can afford a disk array for my PC next year!

Nash [Nash 1993] provides a summary of the status of RAID storage systems as of the summer of 1993. He reports that RAID business worldwide in 1992 was $1.5 billion and is expected to top $2.8 billion in 1993. The top three RAID vendors in 1992 were EMC corporation with $314.9 million, IBM with $209 million, and DEC with $204.9 million.

Nash also reports that currently the price per MB of disk storage for mainframes is about $5.20, but is expected to drop to approximately $1 per MB within four years. He also claims that minicomputer and PC disk drives currently sell for about $3.5/MB and $3.00/MB, respectively, but are expected to drop to $1/MB by 1997. Nash also provides a list of third-party vendors offering RAID systems for different platforms. Platforms included are PCs and networks, Macintosh, UNIX systems, superservers, minicomputers, and mainframes.

Modeling disk I/O can be very easy or very difficult depending upon what level of detail is necessary for your modeling effort. Recall that the total time to complete an I/O operation for a traditional disk drive (not RAID) is the sum of the seek time, latency time, transfer time, controller overhead, RPS miss time for RPS systems, and the queueing or contention time. All of these are easy to compute except the queueing time and the RPS miss time. For modeling systems with no I/O performance problems, that is, with few RPS misses and no queueing, modeling is trivial. Computer systems with I/O problems can often be modeled using queueing network models. If the I/O problems are very serious it might be necessary to use simulation modeling or hybrid modeling. For the hybrid modeling approach simulation is used to model the I/O subsystem in detail to arrive at an accurate average I/O access time. This average access time is then used in a queueing network model as a delay time.

CPU-I/O-Memory Connection

We have been treating the CPU, I/O, and main memory resources somewhat independently; almost as though they really were independent, which they aren't. Of course you must have adequate CPU power to execute a particular workload within a reasonable time frame and with reasonable response time. (No one can do a mainframe job with an original 4.77 MHz IBM PC.) On the other hand, the fastest CPU in the world cannot do much if there is insufficient main memory or insufficient I/O capability.

As Schardt noted earlier, if you don't have enough main memory, you cannot fully utilize the processor. The processor will spend a lot of time waiting for I/O completions.

One of the unmistakable signs of lack of memory is thrashing, that is, paging that is so excessive that almost nothing else is done by the computer. If you have attempted to run large *Mathematica* programs on your PC with insufficient main memory or not enough swapping memory on your hard drive, you have probably experienced this phenomenon. Your hard disk activity light will stay on all the time but there will be almost no indication of new results on your monitor.

There are similar sorts of indications of thrashing that occur on larger machines, of course.

Not enough main memory (or main/expanded on an IBM mainframe or compatible) can also prevent your I/O subsystem from operating properly. Finally, too little main memory sometimes keeps the multiprogramming level so low that the CPU is frequently idle when there is work to be done. The multiprogramming level is low because there is room for only a few programs at a time in main memory. The CPU also could be idle because all the programs in main memory are inactive due to page faults or other I/O requests that are pending.

Naturally, a computer system cannot function well if there is not sufficient I/O capability in the form of disk drives, channels, control units, and I/O caches to handle the I/O required by the application programs. However, for adequate I/O performance there must also be sufficient main memory and sufficient CPU processor power.

Rosenberg's rules mentioned in Chapter 1 provide some guidelines for determining the cause of performance problems. Rosenberg's rules [Rosenberg 1991] are:

> 1. If the CPU is at 100% utilization or less and the required work is being completed on time, everything is okay for now. (But always remember, tomorrow is another day.)
> 2. If the CPU is at 100% busy and all the required work is not completed, you have a problem. Begin looking at the CPU level.
> 3. If the CPU is not 100% busy, and all work is not completed, a problem also exists and the I/O and memory subsystems should be investigated.

Rule 3 conforms to what one would expect; the problem is in the I/O subsystem, the memory subsystem, or both subsystems. Rule 2 is not so obvious. The problem is not necessarily that the CPU is underpowered. By checking to see what the CPU is busy doing you may discover that the CPU is spending too much time on paging activity. As Rosenberg points out, this means there is a memory problem. Checking the CPU activity could also show that the I/O subsystem is causing the problem.

I/O Devices Not Usually Modeled

There are several I/O devices that are not usually explicitly modeled when modeling is used for capacity planning purposes because the devices do not make significant demands on the computer system during "prime time," that is, during the peak periods of the day. These devices include printers, graphic display devices such as computer monitors, and tape drives. Tape drives are usually excluded because they are used primarily as backup devices and are used during off-shift times. It is possible that tape drives need to be modeled as part of the system if there is a great deal of online logging to tape drives. Similarly, for some workstations that do very extensive graphical applications such as CAD, the graphics subsystem must be explicitly modeled. Large printing jobs are usually done off-line so need not be modeled unless the performance problem is in getting the printing done on time.

2.4 Solutions

Solution to Exercise 2.1

We see that $n = \frac{30-20}{20} \times 100 = 50$ percent. so that A is 50% faster than B.

The calculation using **perform** is:

```
In[4]:= perform[20, 30]
Machine A is n% faster than machine B where n = 50.
```

Solution to Exercise 2.2

This exercise is a bit of a red herring. At first glance one would think that a 100 MHz machine running the same code should take exactly half the time that a 50 MHz machine would, that is, in 25 seconds. If everything else were exactly the same, that would be true. Rarely, however, is everything the same. My personal experience is that engineers always make improvements when they produce a new version of any piece of hardware or software. Intel has done more than double the clock speed of their 50 MHz microprocessor to produce a 100 MHz version. They probably have made other hardware improvements as well as improvements in execution algorithms. In addition to this, one would expect Mike Hammer's company to make improvements in the cache and in the memory speed, etc. If you used **cpu** you would obtain:

```
In[6]:= cpu[750000000, 100, 20]
The speed in MIPS is 37.5
The number of clock cycles per instruction, CPI, is
```

```
2.666666667
```

This shows that, if we assume the microprocessor runs at 100 MHz, then the MIPS has jumped to 37.5 and the CPI has dropped to 8/3. These numbers are similar to some of those reported by Intel.

Solution to Exercise 2.3

Using **cpu** we obtain:

```
In[6]:= cpu[750000000, 200, 5]
The speed in MIPS is 150
The number of clock cycles per instruction, CPI, is
1.333333333
```

A 150 MIPS machine with a CPI of 4/3 would be a remarkable machine in 1993 but the Intel 80586 (renamed the Pentium by Intel) approaches some of these performance statistics! The first Pentium-based personal computers were announced by vendors in May 1993. Intel has released two versions of the Pentium, a 60 MHz version and a 66 MHz version. According to [Smith 1993]

> Even the least powerful Pentium PC runs two to three times as fast as a 486. A 60 MHz Pentium PC raced through processor-intensive tests three times as fast as a 486SX/33 and ran Win-Word macros nearly twice as fast as a 486DX2/66.
>
> How does the Pentium deliver its dramatic performance? Four components—two hardware instruction pipelines and two types of caches—are primarily responsible for the Pentium's roughly twofold speed increase over 486 CPUs. No other conventional CPU offers this double dose of pipelines and caches.

It has been suggested that Intel uses the word Pentium to describe the 80586 because "pent" means "five" leading to the suggestion that they should have called it the "Cinco de Micro."

Solution to Exercise 2.4

Note that we use 30% of the reported average seek time of 20 ms. The **simpledisk** solution follows:

```
In[8]:= simpledisk[.3 20, 4800, 150, 8, 2]
```

```
The latency time in milliseconds is 6.25
The transfer time in milliseconds is 0.666667
The access time in milliseconds is 14.9167
```

2.5 References

1. H. Pat Artis, "MVS/ESA: Evolution of the S/390 I/O subsystem," *Enterprise System Journal,* April 1992, 86—93.

2. Kevin Bachus, Patrick Houston, and Elizabeth Longsworth, "Right as RAID," Corporate Computing, May 1993, 61—85.

3. Gordon Bell, "ULTRACOMPUTERS: A teraflop before its time," *CACM*, August 1992, 27—47.

4. Thomas Beretvas, "Paging analysis in an expanded storage environment," *CMG '87 Conference Proceedings,* Computer Measurement Group, 1987, 256—265.

5. Edward I. Cohen, Gary M. King, and James T. Brady, "Storage hierarchies," *IBM Systems Journal*, 28(1), 1989, 62—76.

6. Elizabeth Corcoran, "Thinking Machines: Hillis & Company race toward a teraflops," *Scientific American*, December 1991, 140—141.

7. Peter J. Denning, "RISC architecture," *American Scientist*, January-February 1993, 7—10.

8. Derek L. Eager, John Zahorjan, and Edward D. Lazowska, "Speedup versus efficiency in parallel systems," *IEEE Transactions on Computers,* 38(3), March 1989, 408—423.

9. Horace P. Flatt, "Further results using the overhead model for parallel systems," *IMB Journal of Research and Development*, 36(5/6), September/November 1991, 721—726.

10. Mark B. Friedman, ed, *CMG Transactions, Fall 1991*, Computer Measurement Group. Special issue with selected papers on RAID..

11. John L. Hennessy and David A. Patterson, *Computer Architecture: A Quantitative Approach,* Morgan Kaufmann, San Mateo, CA, 1990.

12. Gilbert E. Houtekamer and H. Pat Artis , *MVS I/O Subsystems: Configuration Management and Performance Analysis*, McGraw-Hill, New York, 1992.

13. Robert H. Johnson, "DASD: IBM direct access storage devices," *CMG'91 Conference Proceedings*, Computer Measurement Group, 1991, 1251—1263.

14. David K. Kahaner and Ulrich Wattenberg, "Japan: a competitive assessment," *IEEE Spectrum*, September 1992, 42—47.

15. Leonard Kleinrock and Jau-Hsiung Huang, "On parallel processing systems: Amdahl's law generalized and some results on optimal design," *IEEE Transactions on Software Engineering,* 18(5), May 1992, 434—447.

16. Elizabeth Lindholm, "Closing the performance gap: as RAID systems mature, vendors are tinkering with the architecture to increase performance, *Datamation*, March 1, 1993, 122—126.

17. Lester Lipsky and C. D. Church, "Applications of a queueing network model for a computer system," *Computing Surveys*, 1977, 205—222.

18. Richard E. Matick, "Impact of memory systems on computer architecture and system organization," I*BM Systems Journal*, 25(3/4), 1986, 274—304.

19. Kim S. Nash, "When it RAIDS, it pours," ComputerWorld, Jun 7, 1993, 49.

20. David A. Patterson, "Expert opinion: Traditional mainframes and supercomputers are losing the battle," *IEEE Spectrum*, January 1992, 34.

21. David A. Patterson, Garth Gibson, Randy H. Katz, "A case for redundant arrays of inexpensive disks (RAID)," *ACM SIGMOD Conference Proceedings*, June 1—3, 1988, 109—116. Reprinted in *CMG Transactions,* Fall 1991.

22. Tom Quinlan, "HP disk array provides secure storage for servers," *InfoWorld*, May 31, 1993, 30.

23. Jerry L. Rosenberg, "More magic and mayhem: formulas, equations and relationships for I/O and storage subsystems," *CMG'91 Proceedings*, Computer Measurement Group, 1991, 1136—1149.

24. Stephen L. Samson, private communication, 1992.

25. Richard M. Schardt, "An MVS tuning approach," *IBM Systems Journal*, 19(1), 1980, 102—119.

26. Gina Smith, "Will the Pentium kill the 496?," PC Computing, May 1993, 116—125.

27. Harold S. Stone, *High-Performance Computer Architecture,* Third Edition, Addison-Wesley, Reading, MA, 1993.

28. Andrew S. Tanenbaum, M. Frans Kaashoek, and Henri E. Bal, "Parallel programming using shared objects and broadcasting," *IEEE Computer*, August 1992, 10—19.

29. Reinhold P. Weicker, "An overview of common benchmarks," *IEEE Computer*, December 1990, 65—75.

Chapter 3 Basic Calculations

A model is a rehearsal for reality, a way of making a trial that minimizes the penalties for error. Playing with a model, a child can practice being in the world. Building a model, a scientist can reduce an object, a system, or a theory to a manageable form. He can watch the behavior of the model, tinker with it—then make predictions about how the plane will fly, how the economy will move, or how a protein chain is constructed.

Horace Freeland Judson

The Search for Solutions

Chance favors the prepared mind.

Louis Pasteur

3.1 Introduction

For all performance calculations we assume some sort of model of the system under study. A model is an abstraction of a system that is easier to manipulate and experiment with than the real system—especially if the system under study does not yet exist. It could be a simple back-of-the-envelope model. However, for more formal modeling studies, computer systems are usually modeled by symbolic mathematical models. (An exception is a detailed benchmark in which real people key in transactions to a real computer system running a real application. Because of the complications and expense of this procedure, it is rarely done.) We usually use a queueing network model when thinking about a computer system. The most difficult part of effective modeling is determining what features of the system must be included and which can safely be left out. Fortunately, using a queueing network model of a computer system helps us solve this key modeling problem. The reason for this is that queueing network models tend to mirror computer systems in a natural way. Such models can then be solved using analytic techniques or by simulation. In this chapter we show that quite a lot can be calculated using simple back-of-the envelope techniques. These are made possible by some queueing network laws including Little's law, the utilization law, the response time law, and the forced flow law. We will illustrate these laws with examples and provide some simple exercise to enable you to test your understanding.

Figure 3.1. Computer System

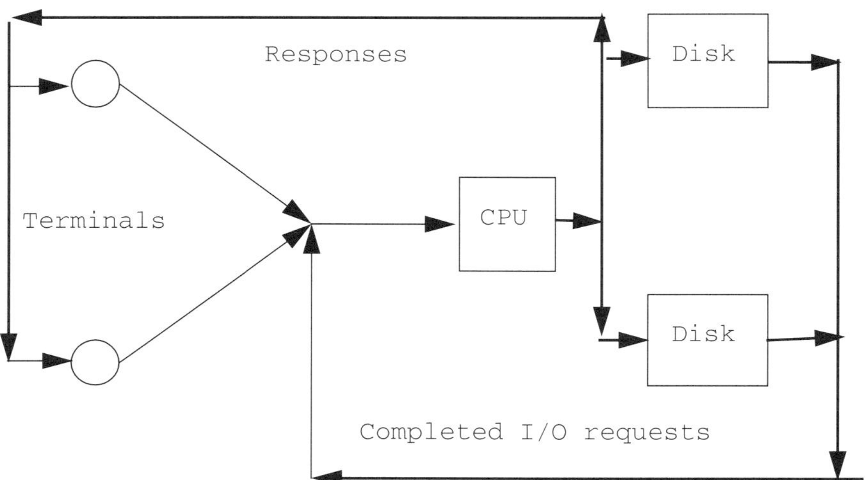

When we think of a computer system a model similar to Figure 3.1 comes to mind. We think of people at terminals making requests for computer service such as entering a customer purchase order, finding the status of a customers account, etc. The request goes to the computer system where there may be a queue for memory before the request is processed. As soon as the request enters main memory and the CPU is available it does some processing of the request until an I/O request is required; this may be due to a page fault (the CPU references an instruction that is not in main memory) or to a request for data. When the I/O request has been processed the CPU continues processing of the original request between I/O requests until the processing is complete and a response is sent back to the user's terminal. This model is a queueing network model, which can be solved using either analytic queueing theory or simulation.

An often overlooked problem with using a model to study a computer system is "falling in love with the model," that is, forgetting that the model is only an approximate representation of the computer system and not the computer system itself. We must always be on guard to ensure that a study utilizing a model does not go beyond the range of validity of the model. The assumptions that are built into the model and whether or not a study extends the parameters of the model beyond the range of applicability must be kept in mind as a modeling study progresses. One should always take the results of a modeling study with a bit of skepticism. Every result should be examined by asking the question, "Is this result reasonable?"

3.1.1 Model Definitions

The queueing network model view of a computer system is that of a collection of interconnected service centers and a set of customers who circulate through the service centers to obtain the service they require as we indicated in Figure 3.1. Thus to specify the model we must define the customer service requirements at each of the service centers, as well as the number of customers and/or their arrival rates. This latter description is called *workload intensity.* Thus workload intensity is a measure of the rate at which work arrives for processing.

Customers are defined in terms of their workload types. Let us first consider single workload class models of computer systems.

3.1.2 Single Workload Class Models

Single workload class models apply to computer systems in which all the users are executing the same application, such as order entry, customer inquiry, electronic mail, etc. For this reason we can treat each customer as being statistically identical, that is, having the same average service requests for each computer resource.

Workload types are defined in terms of how the users interact with the computer system. Some users employ terminals or workstations to communicate with their computer system in an interactive way. The corresponding workload is called a *terminal workload.* Other users run batch jobs, that is, jobs that take a relatively long time to execute. In many cases this type of workload requires special setup procedures such as the mounting of tapes or removable disks. For historical reasons such workloads are called *batch workloads.* (In ancient times such jobs were entered into a computer system by means of a card reader, which read a *batch* of punched cards for each program.) The third kind of workload is called a *transaction workload* and does not correlate quite so closely with the way an actual user utilizes a computer system. Large database systems such as airline reservation systems have transaction workloads, which correspond roughly to computer systems with a very large number of active terminals.

There are two types of parameters for each workload type: parameters that specify the workload intensity and parameters that specify the service requirement of the workload at each of the computer service centers.

We describe the workload intensity for each of the three workload types as follows:

1. The intensity of a terminal workload is specified by two parameters, N, the average number of active terminals (users), and Z, the average think time. The think time is the time between the response to a request and the start of the next request. Neither N nor Z is required to be an integer. Thus a terminal workload could have $N = 23.4$ active users at terminals, on the average, and an average think time of $Z = 10.3$ seconds.

2. The intensity of a batch workload is specified by the parameter N, the average number of active customers (transactions or jobs). Batch workloads have a fixed population. Batch jobs that complete service are thought of as leaving the system to be replaced instantly by a statistically identical waiting job. Thus a batch workload could have an intensity of $N = 6.2$ jobs so that, on the average, 6.2 of these jobs are running on the computer system.

3. A transaction workload intensity is given by λ, the average arrival rate of customers (requests). Thus it has the dimensions of customers divided by time, such as 1,000 inquiries per hour or 50 transactions per second. The population of a transaction workload that is being processed by the computer system varies over time. Customers leave the system upon completing service.

A queueing model with a transaction workload is an open model since there is an infinite stream of arriving and departing customers. When we think of a transaction workload we think of an open system as shown in Figure 3.2 in which requests arrive for processing, circulate about the computer system until the processing is complete, and then leave the system. Conversely, models with batch or terminal workloads are called *closed* models since the customers can be thought of as never leaving the system but as merely recirculating through the system as we showed in Figure 3.1. We treat batch and terminal workloads the same from a modeling point of view; batch workloads are terminal workloads with think time zero. As we will see later, using transaction workloads to model some computer systems can lead to egregious errors. We recommend *fixed throughput workloads* instead, which are discussed in Chapter 4.

There are two types of service centers: queueing and delay. A delay center is often called an *infinite server* service center (IS for short). By this we mean there is always a server available to every arriving customer; no customer must queue for service.(A server is an entity in a service center capable of providing the required service to a customer. Thus a server could be a CPU, an I/O device, etc.) This is approximated in the real world by service facilities which have *enough servers*, that is, sufficiently many servers so that one can always be provided to an arriving customer. We model terminals as delay servers because we assume each user has a terminal and does not need to queue up to use it.

A queueing center is somewhat different and represents the most common service center in a queueing network because customers must compete for service with the other customers. If all the servers at the center are busy, arriving customers join a waiting line to queue (wait) for service. We usually refer to the waiting line as a queue. CPUs and I/O devices are modeled as queueing service centers.

Figure 3.2. Open Computer Model

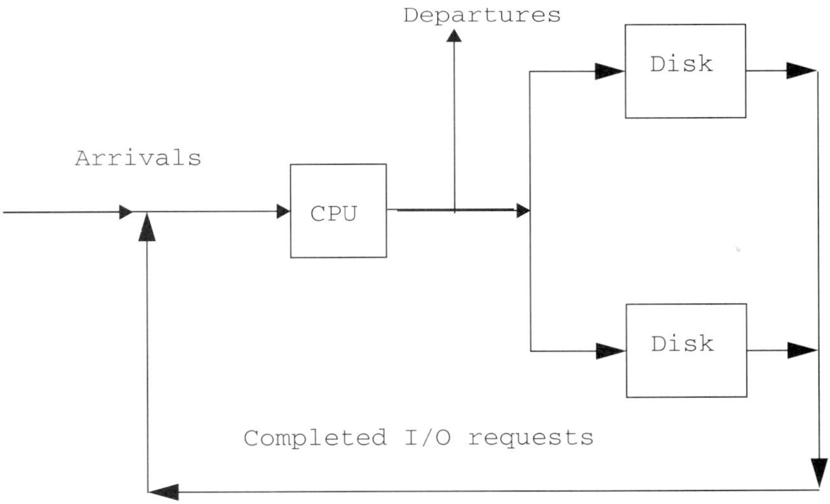

The service demands for a single class model are usually given in terms of D_k, the total service time a customer requires at service center k. (We assume the service centers are numbered 1, 2,...,K.) Sometimes D_k is defined in terms of the average service demand S_k per visit to service center k and the average number of visits V_k that a customer makes to service center k. Then we can write $D_k = V_k \times S_k$. For example, if the service center is the CPU, we may find that the average time a job spends at the CPU on a single visit is 0.02 seconds but that, on the average, 30 visits are required. Then $D_1 = 30 \times 0.02 = 0.6$ seconds.

3.1.3 Multiple Workloads Models

The only difference in nomenclature for models with multiple workload classes is that each workload parameter must be indexed with the workload class number. Thus a terminal class workload has the parameters N_c and Z_c as well as the average service time per visit $S_{c,k}$ and the average number of visits required $V_{c,k}$ for each service center k.

3.2 Basic Queueing Network Theory

A queueing network is a collection of service centers connected together so that the output of any service center can be the input to another. That is, when a customer completes service at one service center the customer may proceed to another service center to receive another type of service.

Figure 3.3. Open Computer Model

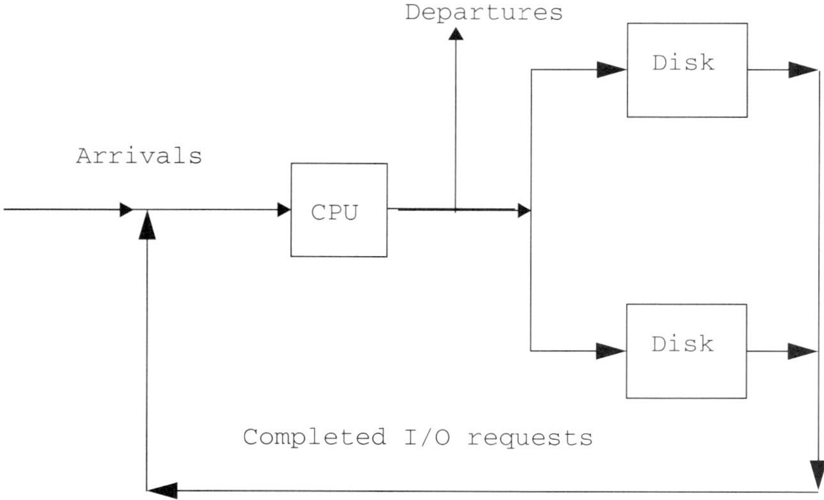

We are following the usual queueing theory terminology of using the word "customer" to refer to a service request. For modeling an open computer system we have in mind a queueing network similar to that in Figure 3.3.In this figure the customers (requests for service) arrive at the computer center where they begin service with a CPU burst. Then the customer goes to one of the I/O devices (disks) to receive some I/O service (perhaps a request for a customer record). Following the I/O service the customer returns to the CPU queue for more CPU service. Eventually the customer will receive the final CPU service and leave the computer system.

We assume that the queueing network representation of a computer system has C customer classes and K service centers. We use the symbol $S_{c,k}$ for the average service time for a class c customer at service center k, that is, for the average time required for a server in service center k to provide the required service to one class c customer. It is the reciprocal of $\mu_{c,k}$, a Greek symbol used to represent the average service rate or the average number of class c customers serviced per unit of time at service center k when the service center is busy. Suppose, for example, that a single workload class computer system has one CPU and we let $k = 1$ for the CPU service center. Then, if the average CPU service requirement is 2 seconds for each customer, we have $S_1 = 2$ seconds and the average service rate for the CPU is $\mu_1 = 0.5$ customers per second.

Some service centers, such as a multiprocessor computer systems with several CPUs, have multiple servers. It is customary to specify the average service time on a per-server basis. Thus, if a multiprocessor system has two CPUs, we specify how long a single processor requires, on the average, to process one customer and designate this number as the average service time. For queueing network models we are not as interested in the average service time of a customer for one visit as we are in the total service demand $D_{c,k} = V_{c,k} \times S_{c,k}$ where $V_{c,k}$ is the average number of visits a class c customer makes to service center k.

Example 3.1

Suppose the performance analysts at Fast Gunn decide to model their computer system as shown in Table 3.1 with one CPU and three I/O devices. They decide to use two workload classes and to number the CPU server as Center 1, with the I/O devices numbered 2, 3, and 4. Both workloads are terminal workloads. Workload 1 has 20 active terminals and a mean think time of 10 seconds, that is, $N_1 = 20$ and $Z_1 = 10$ seconds. Workload 2 has 15 active terminals and a mean think time of 5 seconds, that is, $N_2 = 15$ and $Z_2 = 5$ seconds. The values of the other parameters are shown in Table 3.1.

Note that our statements in the first paragraph of the example plus the table completely define the model. We will demonstrate how to compute predicted performance of the model in Example 3.4

Table 3.1. Data for Example 3.1

c	k	$S_{c,k}$	$V_{c,k}$	$D_{c,k}$
1	1	0.10	5.0	0.500
1	2	0.03	2.5	0.075
1	3	0.04	10.0	0.400
1	4	0.02	20.0	0.400
2	1	0.15	3.0	0.450
2	2	0.03	4.5	0.135
2	3	0.02	8.0	0.160
2	4	0.01	10.0	0.160

3.2.1 Queue Discipline

The queue discipline at a service center is the mechanism for choosing the order in which customers are served if more customers are present than there are servers to serve them. The most common queue discipline is first-come, first-served, abbreviated as FCFS, in which customers are served in order of arrival. This is the queue discipline used in each service line of a fast food restaurant. The antithesis of this queue discipline is last-come, first-served, abbreviated LCFS, in which the last arrival is served first, leaping ahead of earlier arrivals.

Priority queue disciplines also exist in which customers are divided into priority classes and customers are served by class. Customers in the highest priority class get preferential treatment in that they are served before all customers in the next highest priority class, etc. Within a given class the customer preference is FCFS.

There are two basic types of priority queue disciplines; *preemptive* and *nonpreemptive*. In a preemptive priority queueing system, a customer who is receiving service has its service preempted if an arriving customer has a higher priority. The preempted customer returns to the head of its priority class to queue for service. The interrupted service is continued at the interruption point for *preemptive-resume* systems and must be begun from the beginning for *preemptive-repeat* systems. Nonpreemptive systems are called *head-of-the-line* queueing systems, abbreviated HOL.

In recent years a classless queueing discipline called *processor sharing* has been widely used. At a service center with the processor sharing queueing discipline, each customer at the center shares the processing service of the center equally . Thus a processor sharing service center that can service a single customer at the rate of 10 per second services each of 2 customers at the rate of 5 per second or each of 10 customers at the rate of 1 per second.

3.2.2 Queueing Network Performance

In Chapter 1 we mentioned that average response time R and average throughput X are the most common performance metrics for terminal and batch workloads. These same performance metrics are used for queueing networks but both as measurements of system wide performance and measurements of service center performance. In addition we are interested in the average utilization U of each service facility. For any server the average utilization of the device over a time period is the fraction of the time that the server is busy. Thus, if over a period of 10 minutes the CPU is busy 5 minutes, then we have $U = 0.5$ for that period. Sometimes the utilization is given in percentage terms so this utilization would be stated as 50% utilization. Note that the utilization of a service center cannot exceed 100%. We discuss the queueing network performance measurements separately for single workload class models and multiple workload class models.

3.2.2.1 Single Class Performance Measures

The performance measures for a single class model include the system measures shown in Table 3.2 Thus we might have a computer system with an average response time $R = 1.3$ seconds, throughput $X = 3.4$ jobs per second, and number in system $L = 4.42$ jobs.

We also have performance measures to describe the performance of the individual service centers as shown in Table 3.3. For example, if we considered the

Table 3.2. System Performance Measures

System Measure	Description
R	Average system response time
X	Average system throughput
L	Average number in system

Table 3.3. Center Performance Measures

Center Measure	Description
U_k	Average utilization at center
R_k	Average residence (response) time
X_k	Average center throughput
L_k	Average number at center

CPU service center, we might find that the average utilization $U_1 = 0.78$, average response time $R_1 = 0.9$ seconds, average throughput $X_1 = 5.6$ jobs second, and average number at the CPU $L_1 = 5.04$ jobs.

3.2.2.2 Multiple Class Model Performance Measures

Just as for single class models, there are system performance measures and center performance measures for multiple class models.. Thus we may be interested in the average response time for users who are performing order entry as well as for those who are making customer inquiries. In addition we may want to know the breakdown of response time into the CPU portion and the I/O portion so that we can determine where upgrading is most urgently needed. Examples of some of the multiclass performance measures are shown in Example 3.4.

Similarly, we have service center measures of two types: aggregate or total measures and per class measures. Thus we may want to know the total CPU utilization as well as the breakdown of this utilization between the different workloads.

3.3 Queueing Network Laws

3.3.1 Little's Law

The single most profound and useful law of computer performance evaluation (and queueing theory) is called **Little's law** after John D. C. Little who gave the first formal proof in his 1961 paper [Little 1961]. [Little's law is also known as Little's formula and Little's result. I once asked Professor Little which description he preferred. He replied, "I don't care as long as my name is spelled correctly."] Before Little's proof the result had the status of a folk theorem, that is, almost everyone believed the result was true but no one knew how to prove it. The use of Little's law is the most important and useful principle of queueing theory and his paper is the single most quoted paper in the queueing theory literature.

Little's law applies to any system with the following properties:

1. Customers enter and leave the system.

2. The system is in a steady-state condition in the sense that $\lambda_{in} = \lambda_{out}$ where λ_{in} is the average rate that customers enter the system and λ_{out} is the average rate that customers leave the system.

Then, if $X = \lambda_{in} = \lambda_{out}$, L is the average number of customers in the system, and R is the average amount of time each customer spends in the system, we have the relation $L = X \times R$.

Thus Little's law provides a relationship between the three variables L, X, and R. The relationship can be written in two other equivalent forms: $X = L/R$, and $R = L/X$.

3.3.2 Utilization Law

One of the corollaries of Little's law is the **utilization law.** It relates the throughput *X*, the average service time *S*, and the utilization *U* of a service center by the formula $U = X \times S$.

3.3.3 Response Time Law

Consider Figure 3.4. Assume this is a closed single workload class model of an interactive system with *N* active terminals, and a central computer system with one CPU and some I/O devices. Little's law can be applied to the whole system to discover the relation between the throughput *X*, the average think time *Z*, the response time *R*, and the number of terminals *N*. The result is the **response time law**

$$R = \frac{N}{X} - Z.$$

he response time law can to generalized to the multiclass case to yield

$$R_c = \frac{N_c}{X_c} - Z_c.$$

Example 3.2

Suppose the system of Figure 3.4 is a single workload class model having a terminal workload with 45 users, an average think time of 14.5 seconds, and that the system throughput is 3 interactions per second. Then the response time *R* is given by the response time law as $R = 45/3 - 14.5 = 0.5$ seconds. We could perform this calculation in a general form using *Mathematica* as shown in Table 3.4.

Table 3.4. Example 3.2 *Mathematica* Program

Defines the function response	`In[3]:=response[n_,x_,z_]:= r = n/x -z`
Makes the calculation	`In[4]:= response[45,3,14.5]`
The answer	`Out[4]= 0.5`

This completes the example.

Let us consider some further applications of Little's law to the closed model of Figure 3.4. First we consider the CPU by itself, without the queue, to be our system and suppose the average arrival rate to the CPU, including the flow back from the I/O devices, is 60 transactions per second while the average service time per visit of a job to the CPU is 0.01 seconds. Then, by Little's law, the average number of transactions in service at the CPU is $60 \times 0.01 = 0.6$. Now let us consider the application of Little's law to the CPU system consisting of the CPU and the queue for the CPU. Suppose there are 18.6 transactions, on the average, in the CPU system, including those in the queue. Since the average number at the CPU itself is 0.6, this means there are 18 in the queue, on the average. Hence, by Little's law, the average time in the queue is 18/60 = 0.3 seconds. Thus the average total time (queueing plus service) a job spends at the CPU for one pass is 0.3 + 0.01 = 0.31 seconds. We can check this value using Little's law for the system. It yields 18.6 / 60 = 0.31 seconds. (We must have done it right.)

3.3.4 Forced Flow Law

For a single workload class computer system the **forced flow law** says that the throughput of service center *k*, X_k, is given by $X_k = V_k \times X$ where X is the computer system throughput. This means that a computer system is holistic in the sense that the overall throughput of the system determines the throughput through each service center and vice versa.

Figure 3.4. Closed Computer Model

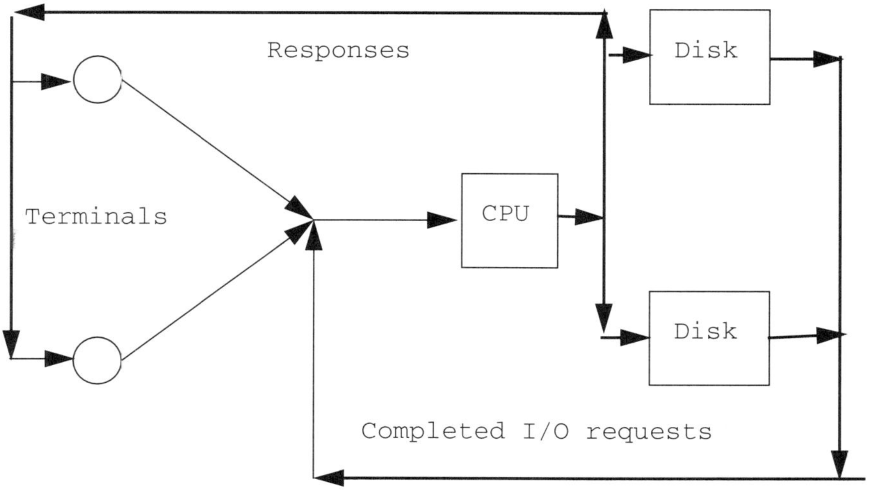

Example 3.3

Suppose Arnold's Armchairs has an interactive computer system (single workload) with the characteristics shown in Table 3.5.

Table 3.5. Data for Example 3.3

Parameter	Description
$N = 10$	There are 10 active terminals
$Z = 18$	Average think time is 18 seconds
$V_{disk} = 20$	Average number of visits to this disk is 20 per interaction
$U_{disk} = 0.25$	Average disk utilization is 25 percent
$S_{disk} = 0.25$	Average disk service time per visit is 0.25 seconds

We make the following calculations:
Since, by the utilization law, $U_{disk} = X_{disk} \times S_{disk}$, we calculate

$$X_{disk} = \frac{U_{disk}}{S_{disk}} = \frac{0.25}{0.025} = 10$$

requests per second.

We can rewrite the forced flow law as $X = X_k/V_k$. Hence, the average system throughput is given by $X = 10/20 = 0.5$ interactions per second. By the response time law we calculate the average response time as $R = 10/0.5 - 18 = 2.0$ seconds.

Example 3.4

You may be wondering what the performance estimates are for the model we described in Example 3.1. Unfortunately, this is a rather complex model to solve. It is one of the models we explain in Chapter 4. However, the *Mathematica* program **Exact** from my book [Allen 1990] (slightly revised) can be used to make the calculations we show here. A revised form of the program called

MultiCentralServer appears in the paper [Allen and Hynes 1991]. It also can be used to make the same calculations. The first line of **Exact** follows:

Exact[Pop_?VectorQ, Think_?VectorQ, Demands_?MatrixQ] :=

We see from this line that the first parameter that must be entered, Pop, is a vector whose components are the number of customers in each class; the next parameter, Think, is a vector of the think times (recall that a batch workload has a think time of zero); and the final parameter, Demands, is an array of service demands. In Example 3.1 we have Pop = {20, 15} because workload class 1 has 20 active terminals and workload class 2 has 15 active terminals. Similarly the entry for the parameter Think is the vector {10, 5}. The service demands of the workloads are given in an array in which row 1 provides the service demands for workload class 1, row 2 the service demands for workload class 2, etc. For this example it is called Demands and is displayed in the *Mathematica* session for Example 3.1 that follows:

```
In[15]:= Think = {10, 5}

Out[15]= {10, 5}

In[16]:= Pop

Out[16]= {20, 15}
In[17]:= MatrixForm[Demands]

Out[17]//MatrixForm= 0.5      0.075    0.4      0.4

                     0.45     0.135    0.16     0.1

In[18]:= Exact[Pop, Think, Demands]


Class# Think    Pop        Resp        TPut
------ ----- ----- --------- ---------
     1    10    20 10.350847   0.98276
     2     5    15  8.278939  1.129608
```

```
Center#     Number        Utiliz
 ------ --------- -----------
      1 16.900123     0.999704
      2  0.291946     0.226204
      3  1.327304     0.573841
      4  1.004985     0.506065
```

The output shows that the CPU is the bottleneck device and is nearly saturated. The second and third disk drives seem to be somewhat heavily utilized according to the performance rules of thumb commonly used.

Exercise 3.1

Consider Example 3.4. Suppose the computer system is upgraded so that the CPU is twice as fast and each I/O device is twice as fast as well. Use **Exact** to calculate the new values for the performance data.

In most of the queueing network algorithms the total service demand at a service center is more important than the service required per visit so we tend to use the service demand D_k at resource k more than we use S_k, the average service time per visit at center k. We also use D with no subscript to be the sum of all the D_k, that is, as the total service time demanded by a job at all resources.

3.4 Bounds and Bottlenecks

One of the key performance concepts used in studying a computer system is the *bottleneck device* or *server*, usually referred to as the *bottleneck*. The name derives from the neck of a bottle, which restricts the flow of liquid. As the workload on a computer system increases some resource of the system eventually becomes overloaded and slows down the flow of work through the computer. The resource could be a CPU, an I/O device, memory, or a lock on a database. When this happens the combination of the saturated resource (server) and a randomly changing demand for that server causes response times and queue lengths to grow dramatically. By *saturated server* we mean a server with a utilization of 1.0 or 100%. A *system is saturated* when at least one of its servers or resources is saturated. The *bottleneck* of a system is the first server to saturate as the load on the system increases. Clearly, this is the server with the largest total service demand.

It is important to note that the bottleneck is workload dependent. That is, different workloads have different bottlenecks for the same computer system. It is part of the folklore that scientific computing jobs are *CPU bound*, while business oriented jobs are *I/O bound*. That is, for scientific workloads such as CAD (computer-aided design), FORTRAN compilations, etc., the CPU is usually the bottleneck. Workloads that are business oriented, such as database management systems, electronic mail, payroll computations, etc., tend to have *I/O* bottlenecks. Of course, one can always find a particular scientific workload that is not CPU bound and a particular business system that is not *I/O* bound, but it is true that different workloads on the same computer system can have dramatically different bottlenecks. Since the workload on many computer systems changes during different periods of the day, so do the bottlenecks. Usually, we are most interested in the bottleneck during the peak (busiest) period of the day.

Example 3.5

Sue Simpson, the lead performance analyst at Sample Systems, measures the performance parameters of a small batch processing computer system. She finds that the CPU has the visit ratio $V_1 = 30$ with $S_1 = 0.04$ seconds, the first I/O device has $V_2 = 10$ and $S_2 = 0.03$ seconds, while the other I/O device has $V_3 = 5$ and $S_3 = 0.04$ seconds. Hence, Sue calculates $D_1 = 1.2$ seconds, $D_2 = 0.3$ seconds, while $D_3 = 0.2$ seconds. She concludes that the bottleneck is the CPU (the system is CPU bound).

Let us now consider some simple bounds for queueing networks.

3.4.1 Bounds for Single Class Networks

For open models the maximum arrival rate λ that the system can process is bounded as follows: $\lambda \le 1/D_{\text{k}}$ where D_{max} is the largest service demand at any service center. The reason for this inequality is that the utilization of every device cannot exceed 1.0 so we must have $U_{\text{k}} = \lambda \times D_{\text{k}} \le 1$ or $\lambda \le 1/D_{\text{max}}$. If the arrival rate exceeds this, the computer system will not be able to keep up with the arrival request stream.

There is a also a lower bound on the average response time given by the best possible performance that can occur. This can occur if there is no queueing for service at any device so that

$$R = \sum_k D_k = D.$$

Unfortunately, there is no upper bound for average response time in open systems.

For closed systems and thus closed workloads, both batch and terminal, there are better bounds than there are with open systems. The same argument we used to show that $1/D_{max}$ is an upper bound on allowed arrival rate for open workloads shows that it is also an upper bound on the throughput X for closed workloads.

For some conditions we can achieve a smaller upper bound than that given by $1/D_{max}$. For example, if there is only one customer in the system, then Little's law implies that $1 = X \times (R+Z)$. Since there is no queueing for service in this case, we have $R = D$ so that $X = 1/(D+Z)$. With more customer arrivals the largest throughput would occur if no customer is delayed by any of the others, that is, if there is no queueing for service. In this case N customers would have the throughput $N/(D+Z)$. Thus, for the general case we have .

$$X \le Min\left(\frac{N}{D+Z}, \frac{1}{D_{max}}\right).$$

There is a lower throughput bound as well, as we now show. By Little's law the throughput is given by $X = N/(R+Z)$ when there are N customers in the system. In the worst possible case, each of the N customers has to queue up behind the other N - 1 customers at each service center so that R, which is the sum of the queueing time plus the service time, is $N \times D$. Therefore, we have $X = N/(ND+Z)$. Since this is the worst possible case, it is a general lower bound so that $N/(ND+Z) \le X$. Combining the last two inequalities we see that

$$\frac{N}{ND+Z} \le X \le Min\left(\frac{N}{D+Z}, \frac{1}{D_{max}}\right).$$

We will now state a useful bound for average response time for batch and terminal workloads. Using the bounds we have derived above and a little algebra we can show that the following upper and lower bounds on the average response time hold:

$$\max[D, N \times D_{max} - Z] \le R \le N \times D.$$

Example 3.6

Consider Example 3.5. For this example $D = D_1 + D_2 + D_3 = 1.7$ seconds and $D_{\max} = D_1 = 1.2$ seconds. If we assume the average number of batch programs in the system (the multiprogramming level) is 5 ($N = 5$), then we have the inequalities

$$0.588235 = \frac{N}{ND+Z} \le X \le \min[\frac{5}{1.7}, \frac{1}{1.2}] = 0.833333.$$

and

$$6 = \max[D, N \times D_{max}] \le R \le N \times D = 8.5.$$

We have shown the brute force back-of-the-envelope solution you could perform with a calculator. The solution using the Mathematica program **bounds** follows:

```
In[4]:= bounds[5, 0, {1.2, 0.3, 0.2}]
Lower bound on throughput is 0.588235
Upper bound on throughput is 0.833333
Lower bound on response time is 6.
Upper bound on response time is 8.5
```

As we ask you to show in Exercise 4.4, the exact answers are $X = 0.831941$ jobs per second and $R = 6.01004$ seconds. At this point you may be thinking, "If I have a Mathematica program that will compute the exact values of X and R for me, what good are the bounds?" The bounds are best used for back-of-the-envelope kinds of calculations when you may be away from your workstation or PC. The bounds are also excellent for validating a model you are developing—especially if it is a simulation model; simulation models are often difficult to validate. (Of course, you could use the exact solution obtained with your *Mathematica* program here, too.) However, if you develop a simulation model, make a long run, and have results for *X* and *R* that do not fall within the bounds, you know there is an error somewhere. Conversely, if the results do fall within the bounds you have some reason for optimism.

Bounds have been developed for multiclass queueing network models but are so difficult to calculate that they are of little practical importance.

3.5 Modeling Study Paradigm

Modeling is an important discipline for studying computer system performance. Most computer performance evaluation experts think of every modeling study as consisting of three phases. They are the *model construction phase* in which a model of the system under study is constructed. As part of this phase tests must be performed to ensure that the model represents the current system with sufficient accuracy for the purpose of the study. The current system is called the baseline system. (The process of determining that the model is a good representation of the current or *baseline system* is called *validation.*)

The second phase is the *evaluation phase* in which the model is modified to represent the system under study after planned changes are made to the hardware, software, and workload. The model is then run to determine the performance parameters of the modified system. Typically the modified model represents a computer system with a more powerful CPU, more memory, more I/O capacity, and (possibly) improved software.

The final phase is the *verification phase* when the actual new system performance is compared to the performance that was predicted during the evaluation phase. This third phase is often not performed but can be very valuable because it helps us improve our modeling techniques.

The most critical part of a modeling study is the setting of clear objectives for the study. Most failed modeling studies fail because the purpose of the study was not clearly understood. We recommend that no modeling study be undertaken without a succinct statement of purpose such as one of the following:

1. To determine whether the improved disk drives we have decided to order should be ordered now or in six months.

2. To decide how much additional memory we need on our current computer system to get us through the next fiscal year.

3. Can the workloads currently running on two model *X* computers be run on one model *Y*?

4. When will computer system *Z* need to be replaced or upgraded?

After the objective of the study is decided upon the model construction phase is begun. The most common case is one in which a current computer system must be modeled. Sometimes the model is of a computer system that does not yet exist, but this is usually the case only for a computer manufacturer who is designing a new line of equipment. We will assume that a model is to be constructed of a current computer system or systems.

As in all modeling, constructing a queueing network model requires that the modeler decide what are the important features of the system modeled that must be included in the model and what features do not have a primary effect and can safely be excluded. The purpose of the model has a big influence here. The model should include only those system resources and workload components that have a primary effect on performance and for which parameter values can be obtained.

An important part of the model construction process is obtaining the required parameter values. This step is called *model parameterization*. The existing computer system is measured to determine the values of the model inputs and the performance with a representative workload, that is, at a representative time. If there is a performance database, the measurements need only be taken from it for a representative period. The baseline model is then constructed with the parameters determined from the measurements. In some cases, as we shall see in Chapter 5, transformations must be made to the original data to generate the model parameters. Some of the parameters, of course, represent the workload. The model is then run to provide performance values such as workload throughput, workload response time, service center utilizations, etc. These model performance values are then compared with the measured performance values to validate the model. Lazowska *et al.* [Lazowska *et al.* 1984] claim that a good analytic queueing network model should be able to predict utilizations within 5% to 10% and response times within 10% to 30%. If the measured values deviate from the predicted values by more than these guidelines, the model must be modified before it is acceptable for prediction.

The first place to look for errors in a model is in the values of the parameters. If nothing can be found wrong with them, then basic changes to the model must be made. More detail may be needed in the representation of the hardware or the workload. Model construction is an iterative process that must continue until a satisfactory model is obtained. Only then can we begin the evaluation phase.

The purpose of the study determines how the evaluation or prediction phase of the study is performed. For example, if the purpose of the study is to determine whether we need improved disk drives now or can wait six months, we would model the system with three different parameterizations: (1) The baseline model with the workload intensity adjusted to that expected in six months, (2) with the new drives installed but with the current workload, and (3) with the new drives installed but with the workload intensity we expect in six months. If the primary performance change of the new drives is that they merely run faster, that is, if there are no major architectural changes in the drives, then the parameter change to represent the new drives is to lower the average service demand for each drive. The first model will estimate the exposure we will suffer if we delay getting the drives. The second will tell us how much improvement to expect if we get the new drives immediately. The third model will give us an estimate of how wonderful it will be in six months if we get the drives now.

The validation phase provides an opportunity to improve our modeling capability. In the disk drive example, if we get the new drives right away, we will have an immediate opportunity to test our model against reality. If the drives are delayed for six months we will be able to test not only our model but our prediction of future workloads.

3.6 Advantages of Queueing Theory Models

G. Scott Graham in his Guest Editor's Overview [Graham 1978] says in part:

> The increasing popularity of queueing network models for computer systems has three bases:
> These models capture the most important features of actual systems, e.g., many independent devices with queues and jobs moving from one device to the next. Experience shows that performance measures are much more sensitive to parameters such as mean service time per job at a device, or mean number of visits per job to a device, than to many of the details of policies and mechanisms throughout the operating system (which are difficult to represent concisely).
>
> The assumptions of the analysis are realistic. General service time distributions can be handled at many devices; load-dependent devices can be modeled; multiple classes of jobs can be accommodated. The algorithms that solve the equations of the model are available as highly efficient queueing network evaluation packages.

Very little can be added to this beautiful statement. The special issue of the *ACM Computing Surveys* in which Graham's statement appears was dedicated to queueing network models of computer system performance; it was published in September 1978 but contains material that is still relevant.

The best known books on queueing theory, especially as the theory can be applied to computer systems, are the two volumes by Kleinrock [Kleinrock 1975, 1976]. These two volumes are distinguished by being clearly written and filled with useful information. Scholars as well as practitioners praise Kleinrock's two volumes.

In this book we will show you how to use queueing network models of computer systems. We will demonstrate how measured data can be used to construct the input parameters for the models and how to overcome the pitfalls that sometimes occur. We will provide *Mathematica* programs to solve the models using both analytic queueing theory as well as simulation and give you an opportunity to experiment with the models.

3.7 Solutions

Solution to Exercise 3.1

We use the *Mathematica* program **Exact** to obtain the output shown. We halved all the service requirements in the array Demands. The other parameters were not changed.

```
In[5]:= MatrixForm[Demands]

Out[5]//MatrixForm= 0.25      0.0375    0.2       0.2

                    0.225     0.0675    0.08      0.05

In[6]:= Pop = {20, 15}

Out[6]= {20, 15}

In[7]:= Think = {10, 5}

Out[7]= {10, 5}
In[8]:= Exact[Pop, Think, Demands]


Class# Think   Pop       Resp        TPut
------ ----- ----- --------- ---------
     1    10    20  2.280246  1.628632
     2     5    15  1.624649  2.264271


Center#     Number       Utiliz
 ------ --------- -----------
      1  5.371382    0.916619
      2  0.269952    0.213912
      3  0.991149    0.506868
      4  0.759842     0.43894
```

We see that there is a tremendous improvement in performance, although the CPU utilization remains high.

3.8 References

1. Arnold O. Allen, *Probability, Statistics, and Queueing Theory with Computer Science Applications,* Second Edition, Academic Press, San Diego, 1990.

2. Arnold O. Allen and Gary Hynes, "Solving a queueing model with *Mathematica,*" *The Mathematica Journal*, 1(3), Winter 1991, 108—112.

3. G. Scott Graham, "Guest editor's overview: Queueing network models of computer system performance," *ACM Computing Surveys*, 10(3), September 1978, 219—224. A special issue devoted to queueing network models of computer system performance.

4. Leonard Kleinrock, *Queueing Systems Volume I: Theory*, John Wiley, New York, 1975.

5. Leonard Kleinrock, *Queueing Systems Volume II: Computer Applications,* John Wiley, New York, 1976.

6. Edward D. Lazowska, John Zahorjan, G. Scott Graham, and Kenneth C. Sevcik, *Quantitative System Performance: Computer System Analysis Using Queueing Network Models*, Prentice-Hall, Englewood Cliffs, NJ, 1984.

7. John D. C. Little, "A proof of the queueing formula: $L = \lambda W$," *Operations Research*, 9(3), 1961, 383—387.

Chapter 4 Analytic Solution Methods

As far as the laws of mathematics refer to reality they are not certain; and as far as they are certain, they do not refer to reality.
Albert Einstein

Sixty minutes of thinking of any kind is bound to lead to confusion and unhappiness.
James Thurber

4.1 Introduction

In Chapter 3 we discussed queueing network models and some of the laws of such models such as Little's law, the utilization law, the response time law, and the forced flow law. We also considered simple bounds analysis. Also discussed were the parameters needed to define a queueing network model and the performance measures that can be calculated for such models. We describe most computer systems under study in terms of queueing network models. Such models can be solved using either analytic solution methods or simulation. In this chapter we will discuss the mean value analysis (MVA) approach to the analytic solution of queueing network models. MVA is a solution technique developed by Reiser and Lavenberg [Reiser 1979, Reiser and Lavenberg 1980]. In Chapter 6 we discuss solutions of queueing network models through simulation.

Although analytic queueing theory is very powerful there are queueing networks that cannot be solved exactly using the theory. In their paper [Baskett *et al.* 1975], a widely quoted paper in analytic queueing theory, Baskett *et al.* generalized the types of networks that can be solved analytically. Multiple customer classes each with different service requirements as well as service time distributions other than exponential are allowed. Open, closed, and mixed networks of queues are also allowed. They allow four types of service centers, each with a different queueing discipline. Before this seminal paper was published most queueing theory was restricted to Jackson networks that allowed only one customer class and required all service times to be exponential. The exponential distribution is a popular one in applied probability because of its nice mathematical properties and because many probability distributions found in the real world are approximately exponential. The networks described by Baskett *et al.* are now known as *BCMP networks*. For these networks efficient solution algorithms are known. Unless we state the contrary we assume that all queueing networks considered in this chapter are BCMP networks.

4.2 Analytic Queueing Theory Network Models

4.2.1 Single Class Models

Strictly speaking, there is a single workload and thus a single class model only if the workload is homogeneous. This means that all the users have the same service demands. This is true if the computer system is used for a single application such as electronic mail or order entry and the users of that application have little variability in their service time requirements. Single class models are sometimes used when the workload is not homogeneous because it is not possible to make the detailed measurements necessary for a multiple class model. In this case the solution will be only approximate but should be more accurate than a simple bounds analysis. Single class models are much easier to solve than multiclass models. In many cases it is possible to solve such a model using back-of-the-envelope techniques and a pocket calculator, especially for open models.

4.2.1.1 Approximation by Open Model

The open, single class model is an approximate model, since there is no actual open, single class computer system. This model is an approximation of a computer system that processes so many transactions that the actual number of terminal users need not be known. A large airline reservation system is such an example. All we need to know to model the system is the average arrival rate and the service demand D_k at each service center. Figure 4.1 indicates how we visualize an open system. We are interested in the maximum throughput possible, which is determined by the bottleneck device, that is, the device that has the maximum service demand $D_{max} = \max\{D_1, D_2, \ldots, D_K\}$. The maximum throughput X_{max} occurs when the bottleneck device is saturated and is given by $X_{max} = 1/D_{max}$. An open system is stable only if $\lambda < X_{max}$ so we make that assumption in our calculations. The calculations for the single, open class model are shown in the Table 4.1. We assume that the average arrival rate as well as the average service demands at the service centers are known. Thus these are the inputs to the model. The outputs or performance measures are what we calculate using the formulas in Table 4.1.

Figure 4.1. Open MVA Model

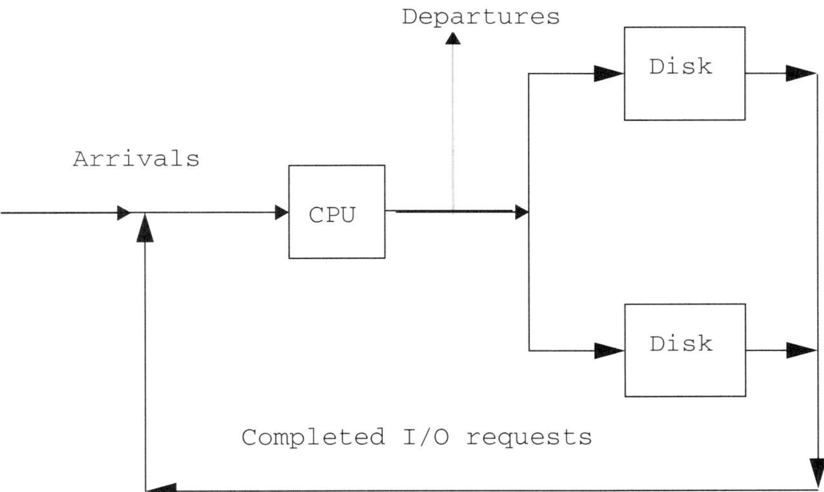

The calculations exhibited in the table can be made using the *Mathematica* program **sopen** from the package **work.m**, which follows Table 4.1.

Table 4.1. Open Model Calculations

Entity	Symbol	Formula
Maximum Throughput	X_{max}	$1/D_{max}$
Center Utilization	U_k	$\lambda \times D_k$
Residence Time	R_k	$D_k/(1-U_k)$
Average Number in Center	L_k	$U_k/(1-U_k)$
System Response Time	R	$\sum R_k$
Average Number in System	L	$\lambda \times R$

```
sopen[ lambda_, v_?VectorQ, s_?VectorQ ] :=
(* single class open queueing model *)
Block[ {n, d, dmax, xmax, u, u1, k},
   d = v s ;
   dmax=Max[d];
   xmax=1/dmax;
   u=lambda*d;
   x=lambda*v;
   numK = Length[v];
   r=d/(1-u);
   l=lambda*r;
   R=Apply[Plus, r];
   L=lambda*R;
   Print[""];
   Print[""];
   Print["The maximum throughput is ",N[xmax, 6]];
   Print["The system throughput is ", N[lambda, 6]];
   Print["The system mean response time is ",N[R, 6]];
   Print["The mean number in the system is ",N[L, 6]];
   Print[ "" ] ;
   Print[ "" ] ;
   Print[
     SequenceForm[
       ColumnForm[ Join[ {"Center#", "------"}, Range[-
numK] ], Right ],
```

```
        ColumnForm[ Join[ {"      Resp", " ----------"},
SetAccuracy[ r, 6 ] ], Right ],
         ColumnForm[ Join[ {"      TPut", " ---------"},
SetAccuracy[ x, 6 ] ], Right ],
         ColumnForm[ Join[ {"    Number", " ---------"},
SetAccuracy[ l, 6 ] ], Right ],
         ColumnForm[ Join[ {" Utiliz", " -------"}, Set-
Accuracy[u, 6]], Right ]]];
    ]
```

Let us consider an example.

Example 4.1

The analysts at Gopher Garbage feel they can model one of their computer systems using the single class open model with three service centers, a CPU and two I/O devices. Their measurements provide the statistics in Table 4.2. Although not shown in the table, they measured the average arrival rate of transactions to be 0.25 transactions per second.

Table 4.2. Input for Example 4.1

Device	V_{device}	S_{device}
CPU	151	0.004
First Disk	80	0.030
Second Disk	70	0.028

The *Mathematica* session used by the Gopher Garbage analysts to produce the statistics for their model follows:

```
In[3]:= <<work.m

In[4]:= v = {151, 80, 70}

Out[4]= {151, 80, 70}

In[5]:= s = {0.004, 0.03, 0.028}
```

```
Out[5]= {0.004, 0.03, 0.028}

In[6]:= sopen[0.25, v, s]

The maximum throughput is 0.416667
The system throughput is 0.25
The system mean response time is 10.5546
The mean number in the system is 2.63864


Center#        Resp       TPut    Number  Utiliz
------ ---------- --------- --------- -------
     1   0.711425      37.75  0.177856   0.151
     2         6.        20.       1.5     0.6
     3   3.843137       17.5  0.960784    0.49
```

It is clear from the output that the first disk is the bottleneck and the cause of the poor performance. The analysts could approximate the effect of adding another disk drive like the first drive and splitting the load over the two drives by using two drives in place of the first drive, each with $V_{disk} = 40$ and $S_{disk} = 0.03$ seconds. We make this change in the following *Mathematica* session:

```
In[5]:= v = {151, 40, 40, 70}

Out[5]= {151, 40, 40, 70}

In[6]:= s = {.004, .03, .03, .028}

Out[6]= {0.004, 0.03, 0.03, 0.028}

In[7]:= sopen[lambda, v, s]


The maximum throughput is 0.510204
The system throughput is 0.25
The system mean response time is 7.98313
The mean number in the system is 1.99578
```

Center#	Resp	TPut	Number	Utiliz
1	0.711425	37.75	0.177856	0.151
2	1.714286	10.	0.428571	0.3
3	1.714286	10.	0.428571	0.3
4	3.843137	17.5	0.960784	0.49

The performance has improved considerably and the new bottleneck appears to be the third disk drive, that is, the one with the mean service time of 0.028 seconds. The effect of further upgrades can easily be tested.

Exercise 4.1

Consider Example 4.1. Suppose that, instead of replacing the first drive with two identical drives, Gopher Garbage decides to replace this drive by one that is twice as fast; that is, by one with a visit ratio of 80 and an average service time of 0.015 seconds. Use **sopen** to make the performance calculations for the upgraded system.

Exercise 4.2

Consider Example 4.1 after the new drive has been added; that is, after the first drive is replaced by two drives. Use **sopen** to estimate the performance that would result for the enhanced system if the third drive is replaced by two drives (one new one), each with a mean service time of 0.028 seconds and with the load split between them.

4.2.1.2 Closed MVA Models

We visualize a closed single class model in Figure 4.2. The N terminals are treated as delay centers. We assume that the CPU is either an exponential server with FCFS queue discipline or a processor sharing (PS) server. By FCFS queueing discipline we mean that customers are served in the order in which they arrive. Processor sharing is a generalization of round-robin in which each customer shares the server equally. The I/O devices are all treated as having the FCFS queue discipline. We assume that the CPU and I/O devices are numbered from 1 to K with the CPU counted as device 1. The MVA algorithm for the performance calculations follows.

Single Class Closed MVA Algorithm. Consider the closed computer system of Figure 4.2. Suppose the mean think time is Z for each of the N active terminals. The CPU has either the FCFS or the processor sharing queue discipline with service demand D_1 given. We are also given the service demands of the I/O devices numbered from 2 to K. We calculate the performance measures as follows:

Step 1 [Initialize] Set $L_k[0] = 0$ for $k = 1, 2, \ldots, K$.

Step 2 [Iterate] For $n = 1, 2, \ldots, N$ calculate

$$R_k[\mathrm{n}] = D_k(1 + L_k[\mathrm{n\text{-}1}]),\ k = 1, 2, \ldots, K,$$

$$R[n] = \sum_{k=1}^{K} R_k[\mathrm{n}],$$

$$X[n] = \frac{n}{R[\mathrm{n}] + Z},$$

$$L_k[\mathrm{n}] = X[\mathrm{n}]\mathrm{R}_k[\mathrm{n}],\ \mathrm{k}=1,2,\ldots,K.$$

Figure 4.2. Closed MVA Model

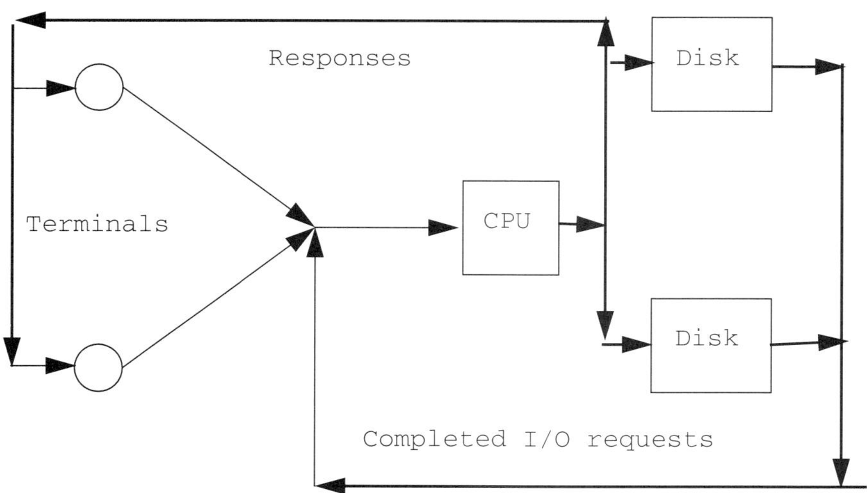

Step 3 [Compute Performance Measures] Set the system throughput to

$$X = X[N].$$

Set response time (turnaround time) to

$$R = R[N].$$

Set the average number of customers (jobs) in the main computer system to $L = X \times R$. Set server utilizations to $U_k = X\,D_k$, k=1,2, ..., K.

We calculated L_k[N] and R_k[N] for each server in the last iteration of Step 2. This algorithm is implemented by the *Mathematica* program **sclosed** which follows:

```
sclosed[N_?IntegerQ, D_?VectorQ, Z_]:=
(* Single class exact closed model *)
Block[{L, r, n, X, u, l, R, K},
K = Length[D];
l=Table[0, {K}];
r=Table[0, {K}];
For[n=1, n<=N, n++, r=D*(l+1); R=Apply[Plus,r]; X=n/
(R+Z);
l=X r; u=X D];
l = X r;
L=X R;
numK = K;
su = u;
  Print[""];
  Print[""]
   Print["The system mean response time is ", R];
   Print["The system mean throughput is ", X];
   Print["The average number in the system is ", L];
   Print[ "" ] ;
   Print[ "" ] ;
   Print[
     SequenceForm[
       ColumnForm[ Join[ {"Center#", "------"},Range[-
numK ] ], Right ],
```

```
        ColumnForm[ Join[ {"     Number", " ---------"},
SetAccuracy[ 1, 6 ] ], Right ],
         ColumnForm[ Join[ {" Utiliz", " -----------"},
SetAccuracy[su, 6]], Right ]]];
         ]
```

The algorithm is actually quite straightforward and intuitive except for the first equation of Step 2, which depends upon the *arrival theorem*, stated by Reiser [Reiser 1981] as follows:

> In a closed queueing network the (stationary) state probabilities at customer arrival epochs are identical to those of the same network in long-term equilibrium with one customer removed.

Like all MVA algorithms, this algorithm depends upon Little's law (discussed in Chapter 3) and the above arrival theorem. The key equation is the first equation of Step 2, $R_k[\text{n}] = D_k(1 + L_k[\text{n-1}])$, which is executed for each service center. By the arrival theorem, when a customer arrives at service station k the customer finds $L_k[\text{n-1}]$ customers already there. Thus the total number of customers requiring service, including the new arrival, is $1 + L_k[\text{n-1}]$. Hence the total time the new customer spends at the center is given by the first equation in Step 2, if we assume we needn't account for the service time that a customer in service has already received. The fact that we need not do this is one of the theorems of MVA! The arrival theorem provides us with a bootstrap technique needed to solve the equation $R_k[\text{n}] = D_k(1 + L_k[\text{n-1}])$ for $n = N$. When n is 1 $L_k[\text{n-1}] = L_k[0] = 0$ so that $R_k[1] = D_k$, which seems very reasonable; when there is only one customer in the system there cannot be a queue for any device so the response time at each device is merely the service demand. The next equation is the assertion that the total response time is the sum of the times spent at the devices. The last two equations are examples of the application of Little's law. The final equation provides the input needed for the first equation of Step 2 for the next iteration and the bootstrap is complete. Step 3 completes the algorithm by observing the performance measures that have been calculated and using the utilization law, a form of Little's law.

Let us illustrate the single class closed MVA model with an example.

Example 4.2

Mellow Memory Makers has an interactive computer system consisting of 50 active terminals connected to a computer system as in Figure 4.2. The performance analysts at MMM find that they can model this system by the queueing model described in the preceding algorithm with one CPU and three disk I/O devices. Their measurements indicate that the average think time is 20 seconds, the mean CPU service demand per interaction is 0.2 seconds, and the mean service demand per interaction on the three I/O devices is 0.03, 0.04, and 0.06 seconds, respectively. The calculations to apply the model can be made with **sclosed** as follows:

```
In[5]:= Demands = {.2, .03, .04, .06}

Out[5]= {0.2, 0.03, 0.04, 0.06}

In[6]:= sclosed[50, Demands, 20]

The system mean response time is 0.523474
The system mean throughput is 2.43623
The average number in the system is 1.2753

Center#        Resp     Number       Utiliz
------ ---------- --------- -----------
     1    0.37695  0.918339    0.487247
     2   0.032312  0.078718    0.073087
     3   0.044215  0.107718    0.097449
     4   0.069997  0.170529    0.146174
```

We see from the output that the throughput is $X = 2.43623$ interactions per second, the mean response time $R = 0.523474$ seconds, the CPU utilization is 0.487247, and the average number of customers (active inquiries) in the computer system is $L = 1.2753$. We also see that the CPU is the bottleneck of the computer system.

Exercise 4.3

Use **sclosed** to find the performance of the Mellow Memory Makers system of Example 4.2 if the CPU is upgraded to twice the current capacity but the I/O devices are retained.

Exercise 4.4

Use **sclosed** to find the exact solution of the computer system described in Examples 3.5 and 3.6. Assume the average population of batch jobs is 5.

4.2.2 Multiclass Models

As we mentioned in Chapter 3, for multiclass models there are performance measures such as service center utilization, throughput, and response time for each individual class. This makes multiclass models more useful than single class models for most computer systems because very few computer systems can be modeled with precision as a single class model. Single class models work best for a computer system that performs only one application. For computer systems having multiple applications with substantially different characteristics, realistic modeling requires a multiclass workload model.

Although multiclass models have a number of advantages over single class models, there are a few disadvantages as well. These include:

1. A great deal more information must be collected to parameterize a multiclass model than a single class model. In some cases it may be difficult to obtain all the information needed from current measurement tools. This may lead to estimates that dilute the accuracy of the multiclass model.

2. As one would expect, multiclass model solution techniques are more difficult to implement and require more computing resources to process than single class models.

These problems, like most worldly problems, can be solved by an infusion of money. Tools for measuring and modeling IBM mainframes running MVS are plentiful and expensive (most but not all of them) but are accurate and relatively easy to use. In fact, the two best known MVS modeling tools, Best/1-MVS from BGS Systems and MAP from Amdahl Corp, can automatically construct a model from RMF data. [RMF (Resource Measurement Facility) is an IBM measurement package.] Best/1-MVS requires the BGS software package CAPTUR/MVS to build the model from RMF and SMF data. [SMF (System Management Facility) is an IBM measurement program designed for capturing accounting information.]

For PCs there are virtually no performance measurement tools other than a few profilers and some CPU and I/O benchmarks such as those supplied by the Norton Utilities, Power, QAPlus, or Checkit.

For some small computers there are not many measurement and modeling tools available. However, most midsize computer systems are supported by their manufacturers and others with both measurement and modeling tools. For example, Digital Equipment Corporation has announced DECperformance Solution V1.0, an integrated product set providing performance and capacity management capabilities for DEC VAX systems running under the VMS operating system. Hewlett-Packard provides an HP Performance Consulting service to help customers with HP 3000 or HP 9000 computer systems solve their performance problems.

4.2.2.1 Multiclass Open Approximation

Just as with single class models, an open multiclass model is an approximation to reality but is fairly easy to implement. Table 4.3 outlines the simple calculations necessary for the multiclass open model. This model assumes that each workload class is a transaction class. The *Mathematica* program **mopen** implements the calculations. We assume that lambda is a vector consisting of the average arrival rates of the classes and Demands is an array that provides the service demands at the service centers by class.

The program **mopen** may not be clear to you if you are not an experienced *Mathematica* programmer, but it does give the correct answers.

Table 4.3. Multiclass Open Model Calculations

Entity	Symbol	Formula
Class c utilization at center k	$U_{c,k}$	$\lambda_c \times D_{c,k}$
Total center k utilization	U_k	$\sum_c \lambda_c \times D_{c,k}$
Time class c customer spends at center k	$R_{c,k}$	$D_{c,k}$ (at delay centers)
Time class c customer spends at center k	$R_{c,k}$	$\frac{D_{c,k}}{1 - U_k}$ (at queueing centers)

Table 4.3. Multiclass Open Model Calculations (Continued)

Entity	Symbol	Formula
Number of class c customers at center k	$L_{c,k}$	$U_{c,k}$ (at delay centers)
Number of class c customers at center k	$L_{c,k}$	$\frac{U_{c,k}}{1 - U_k}$ (at queueing centers)
Number of class c customers in system	L_c	$\sum_k L_{c,k}$
Class c response time	R_c	$\sum_k R_{c,k}$

Let us consider an example.

Example 4.3

The performance analysts at the Zealous Zymurgy brewery feel they can model one of their computer systems using the multiclass open model with the parameters given in Table 4.4.

Table 4.4. Example 4.3 Performance Data

Class		c	k	$D_{c,k}$
1	1.2	1	1	0.20
		1	2	0.08
		1	3	0.10
2	0.8	2	1	0.05
		2	2	0.06
		2	3	0.15
3	0.5	3	1	0.02
		3	2	0.21
		3	3	0.12

The performance analysts enter the data from Table 4.4 into the program **mopen** and obtain their output in the following *Mathematica* session:

```
In[5]:= Demands = {{.2, .08, .1}, {.05, .06, .15},
{.02, .21, .12}}

Out[5]= {{0.2, 0.08, 0.1}, {0.05, 0.06, 0.15}, {0.02,
0.21, 0.12}}

In[6]:= lambda = {1.2, .8, .5}

Out[6]= {1.2, 0.8, 0.5}

In[7]:= mopen[lambda, Demands]

Class#        TPut     Number       Resp
------ ---------- --------- ----------
     1        1.2  0.637286   0.531072
     2        0.8  0.291681   0.364602
     3        0.5  0.239612   0.479225
```

```
Center#     Number      Utiliz
 ------ --------- ----------
      1  0.408451       0.29
      2  0.331558      0.249
      3  0.428571        0.3
```

All times in the output of **mopen** are in seconds. The performance appears to be excellent! Users from each workload class have an average response time that is less than one second. The system is well balanced with each service center almost equally loaded. The second disk drive is loaded slightly higher than the other service centers, making it the bottleneck. We ask you to use **mopen** to determine the effect of replacing the second disk drive by one that is twice as fast.

Exercise 4.5

Consider Example 4.3. Suppose Zealous Zymurgy decides to replace the second disk drive by one that is twice as fast. Assuming the current workload, what are the new values of average response time for each workload class? What would these numbers be if each workload intensity was doubled after the new disk was installed?

4.2.2.2 Exact Closed Multiclass Model

The exact MVA solution algorithm for the closed multiclass model is based on the same ideas as the single class model (Little's law and the arrival theorem) but is much more difficult to explain and to implement. In addition the computational requirements have a combinatorial explosion as the number of classes increases. Increasing the population of a class also increases the computational burden in a dramatic way. I explain the exact MVA algorithm in Section 6.3.2.2 of my book [Allen 1990] and in my article [Allen and Hynes 1991] with Gary Hynes but will refrain from explaining it here because it is beyond the scope of this book. However, we show how to use the *Mathematica* program **Exact**, which is a slightly revised form of the program by that name in my book [Allen 1990]. After considering some examples using **Exact** we consider an approximate MVA algorithm for closed multiclass systems.

Example 4.4

Consider Example 4.2. The solution to the original model using the program **sclosed** required 0.35 seconds on my workstation as we see from the printout below.

```
In[4]:= Demands = {.2, .03, .04, .06}

Out[4]= {0.2, 0.03, 0.04, 0.06}

In[5]:= sclosed[50, Demands, 20]//Timing


The system mean response time is 0.523474
The system mean throughput is 2.43623
The average number in the system is 1.2753

Center#        Resp     Number       Utiliz
 ------ ---------- --------- -----------
      1    0.37695  0.918339    0.487247
      2   0.032312  0.078718    0.073087
      3   0.044215  0.107718    0.097449
      4   0.069997  0.170529    0.146174

Out[5]= {0.35 Second, Null}
```

Suppose we convert to a model with two classes by arbitrarily placing each user into one of two identical terminal classes. Then we solve the model using **Exact** as follows:

```
In[4]:= Demands = {.2, .03, .04, .06}

Out[4]= {0.2, 0.03, 0.04, 0.06}

In[5]:= Demands = {Demands, Demands}

Out[5]= {{0.2, 0.03, 0.04, 0.06}, {0.2, 0.03, 0.04,
0.06}}

In[6]:= Pop = {25, 25}

Out[6]= {25, 25}
```

```
In[7]:= Think = {20, 20}

Out[7]= {20, 20}

In[8]:= Exact[Pop, Think, Demands]//Timing

Class# Think   Pop       Resp       TPut
------ ----- ----- --------- ---------
     1    20    25  0.523474  1.218117
     2    20    25  0.523474  1.218117


Center#    Number       Utiliz
 ------ --------- -----------
      1  0.918339    0.487247
      2  0.078718    0.073087
      3  0.107718    0.097449
      4  0.170529    0.146174
Out[8]= {18.32 Second, Null}
```

We get exactly the same performance statistics as before but it took 18.32 seconds to run the multiclass model compared to only 0.35 seconds for the single class model!

Exercise 4.6

Verify that the output of **Exact** in Example 4.4 does provide the same performance statistics as the output of **sclosed**.

4.2.2.3 Approximate Closed Multiclass Algorithm

The explanation of the approximate MVA algorithm for closed multiclass queueing networks is also beyond the scope of this book but can be found on pages 413–414 of my book [Allen 1990]. It is implemented by the *Mathematica* program **Approx**, which is a slightly modified form of the program by the same name in my book. As can be seen from the first line of the program below, the program expects as input exactly the same inputs as those of **Exact** followed by a number epsilon expressing the size of the error criterion. It is common to use values such as 0.001 for epsilon. The smaller epsilon is, the closer the output of **Approx** is to the solution. Unfortunately, the approximate solution is usually not the same as the exact solution. That is, although the algorithm converges very quickly to a solution, the solution it produces is not usually the exact solution, no matter how small we make epsilon. However, the solution is usually sufficiently close to the exact solution for all practical purposes. Thus the approximate algorithm allows us to model many computer systems that it would not be practical to model using the exact algorithm. Let us consider some examples. We display the first line of **Approx** here so you can see what the inputs are:

```
Approx[ Pop_?VectorQ, Think_?VectorQ, Demands_?Ma-
trixQ, epsilon_Real ] :=
```

Example 4.5

Consider Example 3.4. We show the solutions of that example using **Exact** and **Approx** with an epsilon of 0.001, and **Approx** with an epsilon of 0.000001. Note that the exact solution using **Exact** required 9.45 seconds on my workstation, **Approx** with an epsilon of 0.001 required 1.24 seconds, and **Approx** with an epsilon of 0.000001 took 1.85 seconds. The calculated performance measures from **Approx** changed very little as epsilon was dropped from 0.001 to 0.000001. The differences in output values between **Approx** and **Exact** run from about 2 to 6 percent. This is not as bad as it may first appear because the uncertainty of the values of input data, especially for predicting input values for future time periods, is often larger than that.

```
In[4]:= Pop = {20, 15}

Out[4]= {20, 15}

In[5]:= Think = {10, 5}

Out[5]= {10, 5}

In[6]:= Demands = {{.5, .075, .4, .4}, {.45, .135,
.16, .1}}

Out[6]= {{0.5, 0.075, 0.4, 0.4}, {0.45, 0.135, 0.16,
0.1}}

In[7]:= Exact[Pop, Think, Demands]//Timing
```

```
Class# Think   Pop       Resp      TPut
------ ----- ----- --------- ---------
     1    10    20 10.350847   0.98276
     2     5    15  8.278939  1.129608

Center#    Number      Utiliz
 ------ --------- -----------
      1 16.900123    0.999704
      2  0.291946    0.226204
      3  1.327304    0.573841
      4  1.004985    0.506065

Out[7]= {9.45 Second, Null}

In[8]:= Approx[Pop, Think, Demands, 0.001]//Timing

Class# Think   Pop       Resp      TPut
------ ----- ----- --------- ---------
     1    10    20    10.743     0.964
     2     5    15     8.757      1.09

Center#           number Utilization
 -------------------- -----------
      1     17.453112     0.97275
      2      0.278483    0.219511
      3        1.2268    0.560129
      4      0.948024    0.494708

Out[8]= {1.24 Second, Null}

In[9]:= Approx[Pop, Think, Demands, 0.000001]//Timing

Class# Think   Pop       Resp      TPut
------ ----- ----- --------- ---------
     1    10    20    10.744     0.964
     2     5    15     8.758      1.09
```

```
Center#          number Utilization
 -------------------- -----------
      1       17.454672    0.972699
      2        0.278458    0.219499
      3        1.226191    0.560103
      4        0.947815    0.494687

Out[9]= {1.85 Second, Null}
```

Exercise 4.7

The computer performance analysts at Serene Syrup studied one of their computer systems and found it could be analyzed as a closed system with three workload classes, two terminal and one batch. Tables 4.5 and 4.6 define the inputs to the current model. Find the performance statistics for the computer system using **Exact** and compare the results to the solution using **Approx** with an epsilon of 0.01. Also compare the solution times.

Table 4.5. Input for Exercise 4.7

c	N_c	Think_c
1	5	*20*
2	5	*20*
3	9	*0*

Table 4.6. More Exercise 4.7 Input

c	k	$D_{c,k}$
1	1	0.25
1	2	0.08
1	3	0.12
2	1	0.20
2	2	0.40

Table 4.6. More Exercise 4.7 Input (Continued)

c	k	$D_{c,k}$
2	3	0.60
3	1	0.60
3	2	0.10
3	3	0.12

4.2.2.4 The Approximate MVA Algorithm with Fixed Throughput Classes

There is an approximate MVA algorithm for modeling computer systems that (simultaneously) have both open and closed workload classes. (Recall that transaction workload classes are open although both terminal and batch workloads are closed.) The algorithm for solving mixed multiclass models is presented in my book [Allen 1990] on pages 415—416 with an example of its use. However, we do not recommend the use of this algorithm for reasons that we now elucidate.

As explained in [Allen and Hynes 1991] there are three reasons for using transaction (open) workload classes even though there are no truly open workload classes; open classes are an abstraction or approximation of actual workload classes. Some of the reasons transaction class workloads are sometimes used include:

1. It is much easier to parameterize a transaction class than a terminal or batch class.
2. A mixed class MVA model is easier to solve than a closed multiclass model.
3. It is sometimes very useful to be able to convert a workload class to one in which the throughput is fixed.

The remainder of this section is based on [Allen and Hynes 1991].

The first reason for using transaction workloads is an important reason. Workload models for the baseline system (recall that the baseline system is the system that is originally modeled in a modeling study, that is, it is the current system) are usually derived from measurement data. For both terminal and batch systems it is often difficult to determine the size of the population, that is, the average number of terminals in use or the average number of active batch jobs, directly from the measurement data. In addition, users who project their future workloads often can predict their future volume of work only in terms of throughput required, that is, in terms such as the number of transactions per month or week rather than in the average number of active terminals. It is common practice for modelers in this situation to replace such a workload by a transaction workload with the same throughput and the same service demands as the original measured workload.

The second reason for using transaction workloads is not very important since efficient algorithms for approximating closed models exist. An example is the algorithm we use in **Approx**.

The third reason is important, too; we illustrate it with an example.

While modeling customer systems with queueing network models at the Hewlett-Packard Performance Technology Center we discovered that the use of open (transaction) workloads sometimes causes problems in modeling multiple class workloads. One would expect a closed workload with a small population to be poorly represented as an open class because an open class has an infinite population. This expectation is easy to verify. In addition, we found that in using the approximate MVA mixed multiclass algorithm, significant closed workloads (that is, workloads with high utilization of some resources) represented as an open workload class can cause sizable errors in other classes which must compete for resources at the same priority level. We avoid these problems by using a modified type of closed workload class that we call a *fixed throughput* class. We developed an algorithm that converts a terminal workload or a batch workload into a modified terminal or batch workload with a given throughput. In the case of a terminal workload we use as input the required throughput, the desired mean think time, and the service demands to create a terminal workload that has the desired throughput. We also compute the average number of active terminals required to produce the given throughput. The same algorithm works for a batch class workload because a batch workload can be thought of as a terminal workload with zero think time. For the batch class workload we compute the average number of batch jobs required to generate the required throughput.

We present an example that illustrates difficulties that arise in using transaction workloads in situations in which their use seems appropriate. We also show how fixed throughput classes allow us to obtain satisfactory results. There are cases, of course, in which the use of transaction workloads to represent batch or terminal workloads does produce satisfactory results.

Example 4.6

The analysts at Hooch Distilleries have successfully modeled one of their computer systems using the approximate MVA model with three batch workload classes and three service centers—a CPU and two I/O devices. The service demands are shown in Table 4.7. All times are in seconds

The populations of workload classes A, B, and C are one, two, and one, respectively. Using this information and that from Table 4.7, the analysts at Hooch use the *Mathematica* program **Approx** to obtain the performance results shown in Tables 4.8 and 4.9. All times in the tables are in seconds and throughputs in transactions per second. The Hooch analysts are satisfied that the model values are a good approximation to the measured values of their system. We treat them as identical in this example.

Table 4.7. Example 4.6 Data

c	k	$D_{c,k}$
A	CPU	*300.0*
	I/O 1	*90.0*
	I/O 2	*60.0*
B	CPU	*90.0*
	I/O 1	*0.6*
	I/O 2	*12.0*
C	CPU	*1800.0*
	I/O 1	*18.0*
	I/O 2	*9*

Table 4.8. Output for Example 4.6

c	X_c	Rc
A	0.000751	1330.858
B	0.005565	359.369
C	0.000145	6882.214

Table 4.9. More Output for Example 4.6

k	U_k	L_k
CPU	0.98784	3.803
I/O 1	0.07358	0.074
I/O 2	0.11318	0.122

The Hooch Distilleries manager responsible for the computer installation decided that workload C probably should be removed from the computer system and added to another. She asked the performance analysts to determine how that would change the performance of workload classes A and B. Nue Analyst, the latest addition to the performance staff, was asked to model the current system without workload class C. Nue decided to run **Approx** with workload classes A and B parameterized as before. This approach yielded the performance predictions shown in the **Approx** output from the following *Mathematica* session:

```
In[4]:= Demands = {{300.0, 90.0, 60.0}, {90, 0.6,
12.0}}

Out[4]= {{300., 90., 60.}, {90, 0.6, 12.}}

In[5]:= Think = {0, 0}

Out[5]= {0, 0}
```

```
In[6]:= Pop = {1, 2}

Out[6]= {1, 2}

In[7]:= Approx[Pop, Think, Demands, 0.001]


Class# Think   Pop          Resp          TPut
------ ----- ----- ------------- -----------
     1     0     1    1024.68928    0.000976
     2     0     2    265.496582    0.007533

Center#          number Utilization
 -------------------- -----------
      1        2.741516    0.970747
      2        0.093195    0.092351
      3        0.165289    0.148951
```

Nue is very disappointed with the results. He thought that removing workload class C from the system would greatly improve the performance of the system in processing workload classes A and B but the CPU is still almost saturated while the turnaround times for workload classes A and B are down only 23 percent and 26 percent, respectively. Suddenly he realizes that he has not modeled the workload correctly. The way he modeled the system makes it do more class A and class B work than the original measured system did. To do the same amount of work in the same amount of time the model should have the same throughput rates for each workload class as the measured system. Nue decides to model the modified system with transaction workloads having the same throughputs as the original measured system. He decides to validate this model by modeling the current system with three transaction class workloads—the first having the same throughput and service demands as workload class A, the second the same as workload class B, and the third like workload class C. If the output of this model predicts performance that is close to the measured values, the model is validated. He uses the *Mathematica* program **mopen** in the *Mathematica* session that follows:

```
In[4]:= Demands = {{300, 90, 60}, {90, .6, 12}, {1800,
18, 9}}

Out[4]= {{300, 90, 60}, {90, 0.6, 12}, {1800, 18, 9}}
```

```
In[5]:= Thru = {.00075137, .00556506, .00014529}

Out[5]= {0.00075137, 0.00556506, 0.00014529}

In[6]:= mopen[Thru, Demands]


Class#         TPut     Number             Resp
------ ---------- --------- --------------
     1   0.000751  18.58259    24731.609988
     2   0.005565 41.093631     7384.220603
     3   0.000145 21.420164  147430.410089


Center#     Number      Utiliz
 ------ --------- ----------
       1 80.889351   0.987788
       2  0.079421   0.073578
       3  0.127613   0.113171
```

This output is very different from that in Tables 4.8 and 4.9. The modeled response time for workload class A has increased 1,754 percent, for workload class B by 1,950 percent, and for workload class C by 2,038 percent! The use of transaction workloads will clearly not work here. It is hard to believe that the transaction workload model predicts an average response time for workload class C that is 21.38 times as big as the measured value. The reason for this very large discrepancy is that a workload class with a small finite population is represented in this model as a workload class with an infinite population.

If we now run the *Mathematica* program **Fixed**, requesting the throughputs shown in Table 4.8 with the service demands of Table 4.7, we obtain the output shown:

```
In[4]:= Demands = {{300.0, 90.0, 60.0},{90, 0.6,
12.0}, {1800.0, 18, 9}}

Out[4]= {{300., 90., 60.}, {90, 0.6, 12.}, {1800., 18,
9}}
```

```
In[5]:= MatrixForm[ percent]

Out[5]//MatrixForm= 300.    90.    60.

                    90      0.6    12.

                    1800.   18     9

In[6]:= Think = {0, 0, 0}

Out[6]= {0, 0, 0}

In[7]:= ArrivalRate = {0.000751, 0.005565, 0.000145}

Out[7]= {0.000751, 0.005565, 0.000145}

In[8]:= Fixed[ArrivalRate, {,,}, Think, Demands,
0.001]


  Class#           ArrivR            Pc
-------- -------------- ------------
       1       0.000751      0.993882
       2       0.005565        1.9844
       3       0.000145      0.991998
Class#             Resp            TPut
-------- -------------- -------------
       1    1323.411353        0.000751
       2     356.585718        0.005565
       3    6841.362715        0.000145


  Center#          Number     Utiliz
--------- ------------- ---------
        1       3.773514    0.98715
        2       0.074399  0.073539
        3       0.122366  0.113145
```

It should be clear that **Fixed** generates performance parameters that are almost exactly the same as those in Tables 4.8 and 4.9. Note that the output of **Fixed** has a column for the estimated population of the workload classes. Note, also, that these numbers are very close to the actual sizes of the original populations. It might not be clear to you how to use **Fixed**. To explain how it is used, let us look at the whole program. In spite of the name, the program will calculate the performance statistics for ordinary terminal and batch workload classes as well as fixed workload classes, using the approximation techniques presented in the program **Approx**. **Fixed** was written by Gary Hynes for our joint paper [Allen and Hynes 1991]. Some of the notation is slightly different from that used in this book.

In the first line of the program,

```
Fixed[ Ac_, Nc_,   Z_c_, Dck_, epsilon_Real ] :=
```

each element of the vector Ac is zero for a terminal or batch class but the desired throughput for a fixed class. Since we have only fixed classes for this example we used ArrivalRate, a vector of the desired throughputs, for Ac. Each element of the vector Nc is blank for fixed classes and the actual population of the class for terminal or batch classes. For this example we entered {,,} for Nc because all three classes were considered fixed classes. The input vector Zc has as component *c* the mean think time for the class *c* workload. The component is zero for batch classes and the mean think time for terminal classes. The array Dck is an array such that the element in row *c* and column *k* is the service demand of the class *c* workload at service center *k*. Finally, epsilon is the error criterion. We used an epsilon of 0.001 in this example.

The vector Pc is a bit unusual. If component *c* is a fixed class that component of Pc is the estimate provided by **Fixed** of the population N_c of class *c*. Since all components in our example are fixed class, the final output is composed of these estimates. In general, if class *c* is not a fixed class, component *c* of Pc is X_c, the calculated throughput of class *c* customers. If you see a non-zero number in the column labeled ArrivR in the output, then the corresponding number in the column Pc is the estimate provided by **Fixed** of the population N_c of class *c*. If the number in the column labeled ArrivR is zero, then the number in column Pc is X_c, the calculated throughput of class *c* customers.

In the Mathematica calculations that follow, Nue uses the **Fixed** program to estimate the performance of the current system with workload C removed. He assumes the currently measured throughput rates for workloads A and B.

```
In[5]:= Demands = {{300.0, 90.0, 60.0}, {90.0, 0.6,
12.0}}
```

```
Out[5]= {{300., 90., 60.}, {90., 0.6, 12.}}
In[6]:= MatrixForm[%]

Out[6]//MatrixForm= 300.   90.    60.

                    90.    0.6    12.

In[7]:= Think = {0, 0}

Out[7]= {0, 0}

In[8]:= ArrivalRate = {0.000751, 0.005565}

Out[8]= {0.000751, 0.005565}

In[9]:= Fixed[ArrivalRate, {,}, Think, Demands,
0.001]//Timing

  Class#           ArrivR             Pc
-------- -------------- ------------
       1       0.000751      0.497256
       2       0.005565       0.76552


  Class#             Resp             TPut
-------- -------------- -------------
       1     662.125684       0.000751
       2     137.559796       0.005565


Center#          Number    Utiliz
--------- ------------- ---------
        1      1.073166   0.72615
        2      0.071396  0.070929
        3      0.118214   0.11184

Out[9]= {0.32 Second, Null}
```

The predicted performance values seem very reasonable. Note that the model predicts that 0.497256 class A batch jobs and 0.76552 class B batch jobs must be in the system on the average. This is the end of Example 4.6.

We provide several additional examples of the use of **Fixed** in [Allen and Hynes 1991]. We also discuss an extension of the fixed class algorithm to handle the case in which the analysts request a higher throughput for a workload class than the system is able to deliver. The modification to the algorithm detects this problem and outputs the maximum throughput that the system can deliver.

Exercise 4.8

Consider Example 4.6. Suppose Hooch Distilleries does make the planned change to the system studied in the example and the performance is very close to that predicted by **Fixed**. Use **Fixed** to predict the response time for class A and class B workloads if the throughput for each class increases by 20 percent. Assume the service demands do not change. Use an epsilon of 0.001.

4.2.3 Priority Queueing Systems

In all of our previous models we have assumed that there are no priorities for workload classes, that is, that all are treated the same. However, most actual computer systems do allow some workloads to have priority, that is, to receive preferential treatment over other workload classes. For example, if a computer system has two workload classes, a terminal class that is handling incoming customer telephone orders for products and the other is a batch class handling accounting or billing, it seems reasonable to give the terminal workload class priority over the batch workload class. We will give an example of this.

Every service center in a queueing network has a *queue discipline* or algorithm for determining the order in which arriving customers receive service if there is a conflict, that is, if there is more than one customer at the service center. The most common queue discipline in which there are no priority classes is the first-come, first-served assignment system, abbreviated as FCFS or FIFO (first-in, first-out). Other nonpriority queueing disciplines include last-come, first-served (LCFS or LIFO), and random-selection-for-service (RSS or SIRO). There are also some whimsical queue disciplines that are part of the queueing theory folklore. These include BIFO (biggest-in-first-out), FISH (first-in, still-here), and WINO (whenever-in, never-out). The reader can probably think of others to describe personal experiences with queueing systems.

For priority queueing systems, workloads are divided into priority classes numbered from 1 to n. We assume that the lower the priority class number, the higher the priority, that is, that workloads in priority class i are given preference over workloads in priority class j if $i < j$. That is, workload 1 has the most preferential priority followed by workload 2, etc. Customers within a workload class are served with respect to that class by the FCFS queueing discipline.

There are two basic control policies to resolve the conflict when a customer of class i arrives to find a customer of class j receiving service, where $i < j$. In a *nonpreemptive* priority system, the newly arrived customer waits until the customer in service completes service before beginning service. This type of priority system is called a *head-of-the-line* system, abbreviated HOL. In a preemptive priority system, service for the priority j customer is interrupted and the newly arrived customer begins service. The customer whose service was interrupted returns to the head of the queue for the jth class. As a further refinement, in a *preemptive-resume* priority queueing system, the customer whose service was interrupted begins service at the point of interruption on the next access to the service facility.

Unfortunately, exact calculations cannot be made for networks with workload class priorities. However, widely used approximations do exist. The simplest approximation is the *reduced-work-rate* approximation for preemptive-resume priority systems that have the same priority structure at each service center. It works as follows: The processing power at node k for class c customers is reduced by the proportion of time that the service center is processing higher priority customers. Suppose the service rate of class c customers at service center k is $\mu_{c,k}$. Then the effective service rate at node k for class c jobs is given by

$$\hat{\mu}_{c,k} = \mu_{c,k}\left(1 - \sum_{r=1}^{c-1} U_{r,k}\right).$$

The new effective service rate means that the effective service time is given by

$$\hat{S}_{c,k} = \frac{1}{\hat{\mu}_{c,k}}.$$

Note that all customers are unaffected by lower priority customers so that, in particular, priority class 1 customers have the same effective service rate as the actual full service rate. It is also true that for class 1 workloads the network can be solved exactly.

Let us consider an example.

Example 4.7

A small computer system at Symple Symon Sugar has two workload classes, a terminal class and a batch class with the service demands shown in Table 4.10. Assume the average think time for the terminal workload is 20 seconds. The size of the terminal class is 30 and of the batch class is 5. Let us first calculate the performance using no priority with the *Mathematica* program **Approx** with an epsilon value of 0.001. The *Mathematica* output is shown after Table 4.10.

Table 4.10. Example 4.7 Data

c	k	$D_{c,k}$
1	CPU	0.40
	I/O 1	0.12
	I/O 2	0.12
2	CPU	20.00
	I/O 1	15.00
	I/O 2	15.00

```
In[6]:= Demands = {{.4, .12, .12}, {20, 15, 15}}

Out[6]= {{0.4, 0.12, 0.12}, {20, 15, 15}}

In[7]:= Pop = {35, 5}

Out[7]= {35, 5}

In[8]:= Think = {20, 0}

Out[8]= {20, 0}

In[9]:= Approx[Pop, Think, Demands, 0.001]

Class# Think   Pop          Resp         TPut
------ -----  -----  ------------- -----------
     1    20     35       4.862755    1.407728
     2     0      5     259.982513    0.019232

Center#           number Utilization
 -------------------- -----------
      1      10.268244     0.947733
      2       0.788597     0.457408
```

```
     3          0.788597       0.457408
```

Analysts at Symple Symon are not happy with this result because they want the average response time for their terminal customers to be less than 1.5 seconds. They estimate the performance values for a priority system with the terminal workload given priority one and the batch workload priority two as follows: First they compute the performance values as though the only workload was the terminal workload using **Approx** as shown:

```
In[10]:= Pop = {35}

Out[10]= {35}

In[11]:= Think = {20}

Out[11]= {20}

In[12]:= Demands = Drop[Demands, -1]

Out[12]= {{0.4, 0.12, 0.12}}

In[13]:= Approx[Pop, Think, Demands, 0.001]

Class# Think   Pop          Resp         TPut
------ -----  -----  ------------- -----------
     1    20     35        1.39518     1.6358

Center#          number Utilization
 -------------------- -----------
     1          1.79723     0.654353
     2          0.24256     0.196306
     3          0.24256     0.196306
```

For this call of **Approx** the analysts used the original terminal workload class. The average response time is only 1.39518 seconds and the average throughput is 1.635882 interactions per second compared to 4.862755 seconds and 1.407728 interactions per second without priorities. To compute the performance of the batch class, we compute the effective demands of the batch workload by using the formula

$$V_{c,k} \times \hat{S}_{c,k} = V_{c,k} \times \frac{S_{c,k}}{1 - \sum_{r=1}^{c-1} U_{r,k}} = \frac{D_{c,k}}{1 - \sum_{r=1}^{c-1} U_{r,k}} = \hat{D}_{c,k}.$$

We calculate the performance of the batch workload using **Approx** and the effective demands with the following *Mathematica* session.

```
In[22]:= U = N[U, 6]

Out[22]= {0.654353, 0.196306, 0.196306}

In[23]:= Demands = {20.0, 15.0, 15.0}/(1 - U)

Out[23]= {57.8625, 18.6638, 18.6638}
In[24]:= Approx[{5}, {0}, {Demands}, 0.001]


Class# Think   Pop          Resp          TPut
------ -----  -----  -------------  -----------
     1     0      5     300.734038     0.016626

Center#          number Utilization
 --------------------  -----------
      1        4.174313     0.962021
      2        0.412843     0.310304
      3        0.412843     0.310304
```

This shows that the response time with priorities for the batch class is 300.734038 seconds with a throughput of 0.016626 jobs per second. The computation using the *Mathematica* program **Pri** that calculates the performance statistics for the system with priorities follows:

```
In[27]:= Pop = {35, 5}

Out[27]= {35, 5}

In[28]:= Think = {20, 0}

Out[28]= {20, 0}
```

```
In[29]:= Demands = {{.4, .12, .12},{20, 15, 15}}

Out[29]= {{0.4, 0.12, 0.12}, {20, 15, 15}}

In[34]:= Pri[Pop, Think, Demands, 0.001]

Class# Think       Pop          Resp          TPut
------ ----- --------- ------------- ------------
     1    20        35       1.39518     1.635882
     2     0         5    300.738369     0.016626

Center#        Number    Utiliz
--------- ------------- ---------
        1      5.971677  0.986868
        2      0.655337  0.445692
        3      0.655337  0.445692
```

The output from **Pri** yields average response times of 1.39518 and 300.738369 seconds, respectively, for the response times and 1.635882 and 0.016626, respectively, for the throughputs. These are almost exactly the values we calculated with a more indirect approach. Note that these values are only approximate for two reasons: We used the reduced-work-rate approximation for calculating the priorities and we used the approximate MVA techniques as well.

Exercise 4.9

Consider Example 4.6. Use **Pri** to estimate the performance parameters that would result if the first workload class is given preemptive-resume priority over the second workload class. Use an epsilon value of 0.0001.

4.2.4 Modeling Main Computer Memory

Main memory is one of the most difficult computer resources to model although it is often one of the most critical resources. In many cases it must be modeled indirectly. Since the most important effect that memory has on computer performance is in its effect on concurrency, that is, allowing CPU(s), disk drives, etc., to operate independently, the most common way of modeling memory is through the multiprogramming level (MPL).

The simplest (and first) well-known queueing model of a computer system that explicitly models the multiprogramming level and thus main memory is the *central server model* shown in Figure 4.3. This model was developed by Buzen [Buzen 1971].

Figure 4.3. Central Server Model

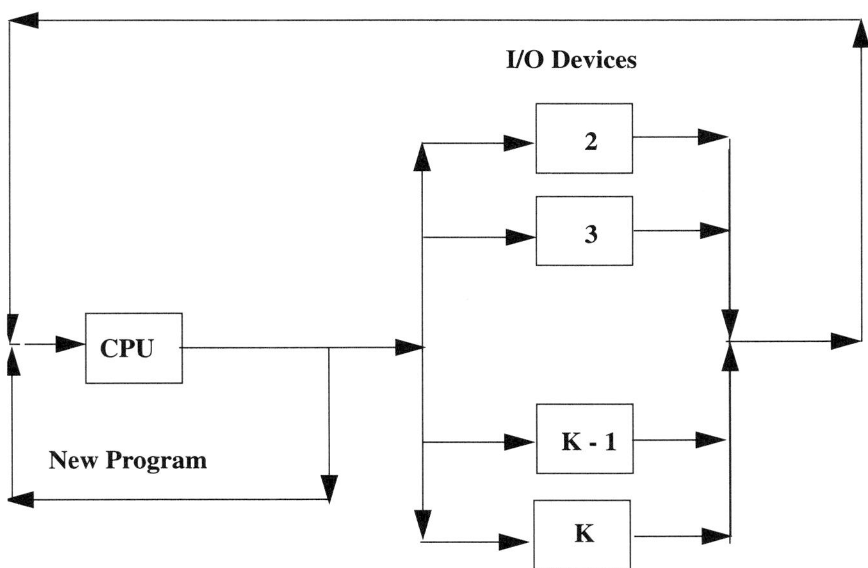

The central server referred to in the title of this model is the CPU. The central server model is closed because it contains a fixed number of programs N (this is also the multiprogramming level, of course). The programs can be thought of as markers or tokens that cycle around the system interminably. Each time a program makes the trip from the CPU directly back to the end of the CPU queue we assume that a program execution has been completed and a new program enters the system. Thus there must be a backlog of jobs ready to enter the computer system at all times. We assume there are K service centers where service center 1 is the CPU. We assume also that the service demand at each center is known. Buzen provided an algorithm called the *convolution algorithm* to calculate the performance statistics of the central server model. We provide a MVA algorithm that is more intuitive and is a modification of the single class closed MVA algorithm we presented in Section 4.2.1.2.

MVA Central Server Algorithm. Consider the central server system of Figure 4.3. Suppose we are given the mean total resource requirement D_k for each of the K service centers and the multiprogramming level N. Then we calculate the performance measures of the system as follows:

Step 1 [Initialize] Set $L_k[0] = 0$ for $k = 1, 2, \ldots, K$.
Step 2 [Iterate] For $n = 1, 2, \ldots, N$ calculate

$$R_k[\mathrm{n}] = D_k(1 + L_k[\mathrm{n}\text{-}1]),\ k = 1, 2, \ldots, K,$$

$$R[n] = \sum_{k=1}^{K} R_k[\mathrm{n}],$$

$$X[n] = \frac{n}{R[\mathrm{n}]},$$

$$L_k[\mathrm{n}] = X[\mathrm{n}]\mathrm{R}_k[\mathrm{n}],\quad \mathrm{k}=1,2,\ldots, K.$$

Step 3 [Compute Performance Measures] Set the system throughput to

$$X = X[N].$$

Set response time (turnaround time) to

$$R = R[N].$$

Set server utilization to

$$U_k = X\,\mathrm{D}_k,\ \mathrm{k}=1,2,\ldots, K.$$

The central server algorithm is valid for the same reasons that the single class closed algorithm is valid. It depends upon repeated applications of Little's law and the arrival theorem. The *Mathematica* program **cent** implements the algorithm. Example 4.8 demonstrates its use.

Table 4.11. Example 4.8 Service Data

k	D_k
CPU	3.5
I/O 1	3.0
I/O 2	2.0
I/O 3	7.5

Table 4.12. Example 4.8 Performance Data

k	U_k	L_k
CPU	0.393	0.553
I/O 1	0.337	0.451
I/O 2	0.225	0.273
I/O 3	0.843	1.724

Example 4.8

The Creative Cryogenics Corporation has a batch computer system that runs only one application. Actually, it is used for other purposes during the day but runs one batch application during the evening hours. Priscilla Pridefull, the chief performance analyst, measures the system and obtains service and performance numbers. All times are in seconds. The average measured turnaround time was 26.69 seconds with an average throughput of 0.11 jobs per second. The service demands are shown in Table 4.11, and the utilizations of and number of customers at each service center are shown in Table 4.12

After verifying that the output of the central server model run with the measured data agreed well with the measured performance, using a multiprogramming level of 3, Priscilla decided to use **cent** to determine what the performance would be if enough additional main memory were obtained to allow a multiprogramming level of 5. (She knows how much memory is needed for the operating systems and other components of the system as well as how much is needed for each copy of the batch program.) Her *Mathematica* run follows the display of the first line from **cent**. Note that, as the first line shows, Priscilla enters the multiprogramming level *N*, and the vector of service demands to execute the program.

```
cent[N_?IntegerQ, D_?VectorQ]:=

In[8]:= Demands

Out[8]= {3.5, 3., 2., 7.5}

In[9]:= cent[5, Demands]

The average response time is 39.2446
The average throughput is 0.127406

Center#     Number      Utiliz
 ------ --------- ----------
      1  0.741785    0.445922
      2  0.585389    0.382219
      3  0.334012    0.254812
      4  3.338814    0.955546
```

We see that the throughput has increased 15.8% to 0.127406 jobs per second (458.66 per hour) while the response time has increased 47% to 39.2446 seconds. We also note that the bottleneck device, the third disk drive, is almost saturated (the utilization is 0.955546).

Priscilla notes that she must do something about the third I/O device. She decides to model the system to see how much improvement would result from splitting the load between the third I/O device and a new identical device. In addition, her users are complaining that it takes too long to run all their batch jobs. They need to get them all done before they must turn the computer system over to the day shift. Priscilla estimates that a throughput of 720 jobs per hour (0.2 jobs per second) will be required within a year to meet the user requirements. She uses the program **Fixed** to decide what multiprogramming level will be needed to be sure of obtaining a throughput of 0.2 jobs per second. **Fixed** computes 8.05661 for the average number of batch jobs needed to obtain a throughput of 0.2 jobs per second, which means that the proper multiprogramming level is probably 8 but could be 9. In the program call of **Fixed**, Priscilla uses braces around 0.15 and 0 (twice), and double braces around the service demands because **Fixed** assumes the service demands are given as an array and that Ac, Nc, and Zc are vectors:

```
In[12]:= Fixed[{0.2}, {0}, {0}, {{3.5, 3.0, 2.0, 3.75,
3.75}}, 0.001]

  Class#          ArrivR           Pc
-------- -------------- ------------
       1            0.2      8.05661

Class#            Resp          TPut
-------- -------------- -------------
       1      40.283041           0.2

Center#          Number    Utiliz
--------- ------------- ---------
        1      1.808472       0.7
        2      1.264335       0.6
        3      0.615689       0.4
        4      2.184056      0.75
        5      2.184056      0.75
```

After running **Fixed** she makes the following calculations using *Mathematica* to check that, with the new I/O device, she needs enough memory to maintain a multiprogramming level of 8 as was predicted by **Fixed**, and that with this multiprogramming level the requirements are met.

```
In[18]:= Demands = {3.5, 3.0, 2.0, 3.75, 3.75}

Out[18]= {3.5, 3., 2., 3.75, 3.75}

In[19]:= cent[8, Demands]

The average response time is 39.9498
The average throughput is 0.200251
```

```
Center#    Number      Utiliz
------ --------- ----------
     1 1.808199     0.70088
     2 1.299528    0.600754
     3 0.636889    0.400503
     4 2.127692    0.750943
     5 2.127692    0.750943
```

Note that Priscilla modeled the new configuration by setting Demands equal to {3.5, 3.0, 2.0, 3.75, 3.75} to account for the new I/O device. For multiprogramming level 8 the throughput exceeds 0.2 jobs per second.

Note, also, that the central server model does not model the CPU and I/O overhead needed to manage memory directly. (Analysts sometimes correct for this by adding a little to the CPU service demand.) In spite of this, the central server model can be used to model some fairly complex systems. For example, in their book [Ferrari, Serazzi, and Zeigner 1983] Ferrari *et al.* used the central server model to find the optimal multiprogramming level in a large mainframe virtual memory system, to improve a virtual memory system configuration, for bottleneck forecasting for a real-time application, and for other studies.

Exercise 4.10

For the final system modeled by Priscilla Pridefull at Creative Cryogenics the third and fourth I/O devices are still the bottlenecks of the system. Suppose the two new I/O devices are replaced by faster I/O devices so that the new average service demands on them are 2.5 seconds. Suppose, also, that enough memory is added so that the multiprogramming level can be increased to 10. Use **cent** to calculate the average throughput and response time of the system. Assume the system will be run at multiprogramming level 10 until all the jobs are completed.

Although the central server model has been used extensively it has two major flaws. The first flaw is that it models only batch workloads and only one of them at a time. That is, it cannot be used to model terminal workloads at all and it cannot be used to model more than one batch workload at a time. The other flaw is that it assumes a fixed multiprogramming level although most computer systems have a fluctuating value for this variable. In the next model we show how to adapt the central server model so that it can model a terminal or a batch workload with a multiprogramming level that changes over time. We need only assume that there is a maximum possible multiprogramming level m.

Since a batch computer system can be viewed as a terminal system with think time zero, we imagine the closed system of Figure 4.2 as a system with N terminals or workstations all connected to a central computer system. We assume that the computer system has a fluctuating multiprogramming level with a maximum value m. If a request for service arrives at the central computer system when there are already m requests in process the request must join a queue to wait for entry into main memory. (We assume that the number of terminals N is larger than m.) The response time for a request is lowest when there are no other requests being processed and is largest when there are N requests either in process or queued up to enter the main memory of the central computer system. A computer system with terminals connected to a central computer with an upper limit on the multiprocessing level (the usual case) is not a BCMP queueing network. The non-BCMP model for this system is created in two steps. In the first step the entire central computer system, that is, everything but the terminals, is replaced by a *flow equivalent server* (FESC). This FESC can be thought of as a black box that when given the system workload as input responds with the same throughput and response time as the real system. The FESC is a *load-dependent* server, that is, the throughput and response time at any time depends upon the number of requests in the FESC. We create the FESC by computing the throughput for the central system considered as a central server model with multiprogramming level 1, 2, 3,..., m. The second step in the modeling process is to replace the central computer system in Figure 4.2 by the FESC as shown in Figure 4.4. The algorithm to make the calculations is rather complex so we will not explain it completely here. (It is Algorithm 6.3.3 in my book [Allen 1990].) However, the *Mathematica* program **online** in the *Mathematica* package **work.m** implements the algorithm. The inputs to **online** are m, the maximum multiprogramming level, *D*emands, the vector of demands for the K service centers, N, the number of terminals, and T, the average think time. The outputs of **online** are the average throughput, the average response time, the average number of requests from the terminals that are in process, the vector of probabilities that there are 0, 1, ..., m requests in the central computer system, the average number in the central computer system, the average time there, the average number in the queue to enter the central computer system (remember, no more than m can be there), the average time in the queue, and the vector of utilizations of the service centers.

Let us consider an example of the use of this model.

Figure 4.4. FESC Form of Central Server Model

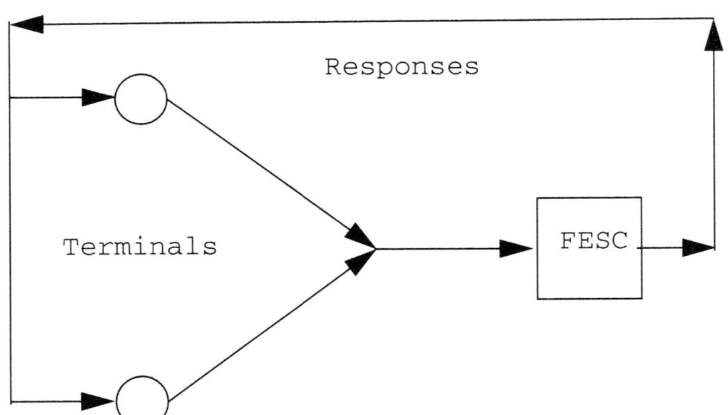

Example 4.9

Meridian Mappers wants to connect their 30 personal computers together by a LAN with a powerful file server; the server can be modeled with one CPU and two I/O devices. Their estimates of the service demands their personal computers will make on the file server are 0.1, 0.2, and 0.25 seconds, respectively, for the CPU, I/O device 1, and I/O device 2. Their average think time is estimated to be 20 seconds and the maximum multiprogramming level that can be achieved by the file server is 5. They hope that this system will provide an average response time that is less than 1 second with an average throughput of at least 1 interaction per second. Their modeling of it with **online** follows:

```
In[12]:= Demands = {.1, .2, .25}
Out[12]= {0.1, 0.2, 0.25}

In[15]:= online[5, Demands, 30, 20]

The average number of requests in process is 1.11835
The average system throughput is 1.44408
The average system response time is 0.774439
The average number in main memory is 1.10942
```

```
Center#      Utiliz
 ------ ----------
      1   0.144408
      2   0.288816
      3   0.361021
```

Thus the requirements of Meridian Mappers would be met according to the model.

Exercise 4.11

Suppose Meridian Mappers of Example 4.9 decides to consider a file server that is half as fast but has I/O devices that are twice as fast, that is, that Demands = {0.2, 0.1, 0.125}, but that will support a maximum multiprogramming level of 10. Use **online** to estimate the performance.

At this point you may be thinking: "You have shown how to model memory in a computer system with either a single batch workload or a single terminal workload, although the latter was a bit complicated. Can memory be modeled in a multiclass workload model?" My answer is a resounding, "Yes, but...." There is no exact model for modeling memory in a computer system with multiple workload classes. However, comprehensive (and expensive) modeling packages such as Best/1 MVS and MAP do model such systems. The bad news about this is that the models are very complex as well as proprietary. At the Hewlett-Packard Performance Technology Center, Gary Hynes has added the capability of modeling memory in multiclass computer systems with hundreds of lines of C++ code. In principle I could translate the code to *Mathematica*, but in practice I cannot. There is no easy way to build a queueing model that can model memory in a multiclass computer system but you can buy a package that will do so. Calaway [Calaway 1991] mentioned that he modeled memory with Best/1 MVS but was unable to do so with the simulation package SNAP/SHOT. Some of his comments follow:

> It should be noted that SNAP/SHOT does not model memory capacity and therefore assumes unlimited memory. Best/1 does model memory, and one scenario was run with a Model J upgrade and increased memory (both central and expanded storage) to determine what effect memory would have on response time and CPU busy time. The response time did not

> change, and the CPU busy went from 73.1 to 72.6 (a difference of 0.5 percent) at the low end and from 93.2 to 93.0 (a difference of 0.2) at the high end. See Figure. This would indicate that our system was not memory constrained.

Clearly it is very useful to be able to model memory. Although SNAP/SHOT does not have this capability, it is possible to model memory using simulation. In fact, simulation is the technique used at the Performance Technology Center to validate our analytic queueing theory model of memory.

4.3 Solutions

Solution to Exercise 4.1

We made the calculations with the following *Mathematica* session:

```
In[5]:= v = {151, 80, 70}

Out[5]= {151, 80, 70}

In[6]:= s = {0.004, 0.015, 0.028}

Out[6]= {0.004, 0.015, 0.028}

In[7]:= sopen[0.25, v, s]


The maximum throughput is 0.510204
The system throughput is 0.25
The system mean response time is 6.26885
The mean number in the system is 1.56721


Center#        Resp       TPut     Number  Utiliz
 ------ ---------- --------- --------- -------
      1   0.711425     37.75  0.177856   0.151
      2   1.714286       20.  0.428571     0.3
      3   3.843137      17.5  0.960784    0.49
```

This output shows that better performance results from replacing the slow disk with a fast disk than with adding a new slow disk and splitting the load between the two. This result is actually a well known result from queueing theory.

Solution to Exercise 4.2

The solution was found from the following *Mathematica* session:

```
In[8]:= lambda = .25

Out[8]= 0.25

In[9]:= v = {151, 40, 40, 35, 35}

Out[9]= {151, 40, 40, 35, 35}

In[10]:= s = {.004, .03, .03, .028, .028}

Out[10]= {0.004, 0.03, 0.03, 0.028, 0.028}

In[11]:= sopen[lambda, v, s]

The maximum throughput is 0.833333
The system throughput is 0.25
The system mean response time is 6.73602
The mean number in the system is 1.68401


Center#        Resp       TPut     Number  Utiliz
 ------ ---------- --------- --------- -------
      1   0.711425     37.75  0.177856   0.151
      2   1.714286       10.  0.428571     0.3
      3   1.714286       10.  0.428571     0.3
      4   1.298013      8.75  0.324503   0.245
      5   1.298013      8.75  0.324503   0.245
```

Adding another drive has certainly improved the performance but the performance of this system is not as good as that of the system in Exercise 4.1.

Solution to Exercise 4.3

The *Mathematica* solution using **sclosed** follows.

```
In[9]:= Demands

Out[9]= {0.1, 0.03, 0.04, 0.06}

In[10]:= sclosed[50, Demands, 20]


The system mean response time is 0.278343
The system mean throughput is 2.46568
The average number in the system is 0.686306

Center#       Resp     Number      Utiliz
 ------ ---------- --------- -----------
      1   0.131598  0.324479    0.246568
      2   0.032341  0.079742    0.073971
      3   0.044270  0.109156    0.098627
      4   0.070134  0.172928    0.147941
```

As the output shows, the average response time has dropped from 0.523474 seconds to 0.278343 seconds, and the number of interactions in process has dropped from 1.2753 to 0.686306, both of which are significant improvements, although the throughput has increased only from 2.43623 interactions per second to 2.46568 interactions per second, a very minor improvement.

Solution to Exercise 4.4

The *Mathematica* solution using **sclosed** follows:

```
In[8]:= Demands

Out[8]= {1.2, 0.3, 0.2}
In[9]:= sclosed[5, Demands, 0]


The system mean response time is 6.01004
The system mean throughput is 0.831941
The average number in the system is 5.0
```

```
Center#        Resp     Number       Utiliz
------ ---------- --------- -----------
     1   5.373012  4.470026     0.998329
     2   0.397604  0.330783     0.249582
     3   0.239428   0.19919     0.166388
```

Thus the system mean response time is slightly larger than the lower bound we calculated in Example 3.6, and the system mean throughput is about halfway between the lower and upper bounds.

Solution to Exercise 4.5

For the first part of the exercise we cut the demand at service center 3 (the second disk drive) to half its original value and apply the program **mopen** as follows:

```
In[12]:= MatrixForm[Demands]

Out[12]//MatrixForm= 0.2      0.08     0.05

                     0.05     0.06     0.075

                     0.02     0.21     0.06

In[13]:= mopen[lambda, Demands]


Class#        TPut     Number       Resp
------ ---------- --------- ----------
     1         1.2  0.536446   0.447038
     2         0.8  0.190841   0.238551
     3         0.5  0.189192   0.378384


Center#     Number      Utiliz
------ --------- ----------
     1  0.408451        0.29
     2  0.331558       0.249
     3  0.176471        0.15
```

The new average response time for each class is 0.447038 seconds (0.531), 0.238551 seconds (0.365), and 0.378384 seconds (0.479), respectively, where the number in parentheses is the value with the slower drive. The improvements are significant but not spectacular.

The performance calculation with the new drive but doubled workload intensities follows:

```
In[15]:= lambda = 2 lambda
Out[15]= {2.4, 1.6, 1.}

In[16]:= mopen[lambda, Demands]

Class#        TPut    Number       Resp
------ ---------- --------- ----------
     1        2.4  1.696756   0.706982
     2        1.6   0.55314   0.345712
     3         1.   0.55166    0.55166

Center#     Number     Utiliz
 ------ --------- ----------
      1  1.380952       0.58
      2  0.992032      0.498
      3  0.428571        0.3
```

We see that the new average response times for the three classes are 0.706982 seconds, 0.345712 seconds, and 0.55166 seconds, respectively. We get excellent response times with twice the load. Perhaps the system is overconfigured!

Solution to Exercise 4.6

The output of **Exact** follows:

```
In[6]:= Demands = {Demands, Demands}

Out[6]= {{0.2, 0.03, 0.04, 0.06}, {0.2, 0.03, 0.04,
0.06}}
In[7]:= Pop = {25, 25}

Out[7]= {25, 25}

In[8]:= Think = {20, 20}
```

```
Out[8]= {20, 20}

In[11]:= Exact[Pop, Think, Demands]//Timing


Class# Think   Pop      Resp       TPut
------ ----- ----- --------- ---------
     1    20    25  0.523474  1.218117
     2    20    25  0.523474  1.218117

Center#     Number      Utiliz
 ------ --------- -----------
      1  0.918339    0.487247
      2  0.078718    0.073087
      3  0.107718    0.097449
      4  0.170529    0.146174

Out[11]= {18.14 Second, Null}
```

The output of **sclosed** follows:

```
In[5]:= sclosed[50, Demands, 20]//Timing

The system mean response time is 0.523474
The system mean throughput is 2.43623
The average number in the system is 1.2753

Center#        Resp    Number      Utiliz
 ------ ---------- --------- -----------
      1    0.37695  0.918339    0.487247
      2   0.032312  0.078718    0.073087
      3   0.044215  0.107718    0.097449
      4   0.069997  0.170529    0.146174

Out[5]= {0.35 Second, Null}
```

The last two columns in the output of each program are identical. These represent the total number of customers and the total utilization, respectively, at the service centers. **sclosed** also provides the residence (response) time at each of the service centers. We do not provide this information as output in **Exact** because it is not very meaningful for a multiclass model (OK, I know you may think that the performance statistics are not exactly the same with this left out, and you are probably right). **sclosed** prints out the average response time, which is 0.523474. This agrees with the average response time of each class in the output of **Exact**. **sclosed** also provides the average throughput, 2.43623 customers per second. In the output of **Exact** we give two numbers for this, one for each class. These numbers are both 1.21812 so their sum is 2.43624. The third number in the output of **sclosed** is 1.2753, the total number of customers in the system. This agrees with the sum of the elements of the next-to-last column in the output from both **sclosed** and **Exact**.

Solution to Exercise 4.7

The *Mathematica* solution follows:

```
In[4]:= Pop = {5, 5, 9}
Out[4]= {5, 5, 9}

In[5]:= Think = {20, 20, 0}

Out[5]= {20, 20, 0}

In[6]:= Demands = {{.25, .08, .12}, {.2, .4, .6}, {.6,
.1, .12}}

Out[6]= {{0.25, 0.08, 0.12}, {0.2, 0.4, 0.6}, {0.6,
0.1, 0.12}}

In[7]:= Exact[Pop, Think, Demands]//Timing

Class# Think   Pop       Resp        TPut
------ ----- ----- --------- ---------
     1    20     5  2.888963  0.218446
     2    20     5  3.481916   0.21293
     3     0     9  5.981389  1.504667

Center#     Number       Utiliz
 ------ --------- -----------
      1  9.542639    0.999998
      2  0.336092    0.253114
      3  0.493754    0.334531

Out[7]= {11.61 Second, Null}
```

```
In[8]:= Approx[Pop, Think, Demands, 0.01]//Timing

Class# Think    Pop       Resp       TPut
------ -----  -----  ---------  ---------
     1    20      5      2.894      0.218
     2    20      5      3.488      0.213
     3     0      9       6.07      1.483

Center#          number Utilization
 -------------------- -----------
      1        9.561966     0.98677
      2        0.328501    0.250888
      3        0.484095    0.331853

Out[8]:= (0.54 Second, Null)
```

The solution using **Approx** is accurate enough for most practical purposes and was generated in much less time.

Solution to Exercise 4.8

The *Mathematica* calculations follow:

```
In[15]:= Demands

Out[15]= {{300, 90, 60}, {90, 0.6, 12}}

In[16]:= ArrivalRate

Out[16]= {0.00075137, 0.00556506}

In[17]:= ArrivalRate = 1.2 %

Out[17]= {0.000901644, 0.00667807}

In[18]:= Think

Out[18]= {0, 0}

In[19]:= Fixed[ArrivalRate, {,}, Think, Demands,
0.001]//Timing
```

```
Class#           ArrivR            Pc
--------  --------------  ------------
       1     0.000901644       0.65881
       2      0.00667807        1.0057

  Class#             Resp            TPut
--------  --------------  -------------
       1      730.676806       0.000902
       2      150.597947       0.006678

  Center#          Number     Utiliz
---------  -------------  ---------
        1        1.435106    0.87152
        2        0.085833   0.085155
        3        0.143576   0.134236
```

From the output we see that R_A = 730.677 seconds and R_B = 150.598 seconds. Thus the response time for workload class A has increased by only 10.35 percent and that of workload B by 9.46 percent. The CPU is the bottleneck and has reached a utilization of 0.87152.

Solution to Exercise 4.9

The *Mathematica* session that provides the answers follows:

```
In[5]:= Pop = {20, 15}

Out[5]= {20, 15}

In[6]:= Think  = {10, 5}

Out[6]= {10, 5}

In[7]:= Demands = {{.5, .075, .4, .4}, {.45, .135,
.16, .1}}

Out[7]= {{0.5, 0.075, 0.4, 0.4}, {0.45, 0.135, 0.16,
0.1}}

In[8]:= Pri[Pop, Think, Demands, 0.0001]
```

```
Class# Think       Pop          Resp         TPut
------ ----- --------- ------------- ------------
     1    10        20      3.569473     1.473897
     2     5        15     21.162786     0.573333

Center#          Number    Utiliz
--------- ------------- ---------
        1     14.052747  0.994948
        2      0.218225  0.187942
        3      1.622588  0.681292
        4      1.500807  0.646892
```

We see that the performance of the first workload class improves considerably. The average response time drops from 10.35 seconds to 3.569473 seconds while the average throughput increases from 0.98276 interactions per second to 1.473897 interactions per second. This improvement for the first workload class leads to poorer performance for the second workload class for which the average response time increases from 8.18 to 21.16 seconds, while the average throughput declines from 1.13 interactions per second to 0.573333 interactions per second.

Solution to Exercise 4.10

The *Mathematica* solution follows.

```
In[7]:= Demands = {3.5, 3.0, 2.0, 2.5, 2.5}

Out[7]= {3.5, 3., 2., 2.5, 2.5}
In[8]:= cent[10, Demands]


The average response time is 39.8201
The average throughput is 0.25113

Center#    Number     Utiliz
 ------ --------- ----------
      1  3.704827   0.878954
      2  2.343416   0.753389
      3   0.95452   0.502259
      4  1.498619   0.627824
      5  1.498619   0.627824
```

Solution to Exercise 4.11

The *Mathematica* solution follows:

```
In[9]:= Demands

Out[9]= {0.2, 0.1, 0.125}

In[10]:= online[10, Demands, 30, 20]

The average number of requests  in process is 0.796053
The average system throughput is 1.4602
The average system response time is 0.545168
The average number in main memory is 0.79605


Center#     Utiliz
------ ----------
     1   0.292039
     2    0.14602
     3   0.182525
```

4.4 References

1. Arnold O. Allen, *Probability, Statistics, and Queueing Theory with Computer Science Applications,* Second Edition, Academic Press, San Diego, 1990.

2. Arnold O. Allen and Gary Hynes, "Solving a queueing model with Mathematica," *Mathematica Journal*, 1(3), Winter 1991, 108—112.

3. Arnold O. Allen and Gary Hynes, "Approximate MVA solutions with fixed throughput classes," *CMG Transactions* (71), Winter 1991, 29—37.

4. Forest Baskett, K. Mani Chandy, Richard R. Muntz, and Fernando G. Palacios, "Open, closed, and mixed networks of queues with different classes of customers," *JACM*, 22(2), April 1975, 248—260.

5. Jeffrey P. Buzen, "Queueing network models of multiprogramming," Ph.D. dissertation, Division of Engineering and Applied Physics, Harvard University, Cambridge, MA, May 1971.

6. James D. Calaway, "SNAP/SHOT VS BEST/1," *Technical Support*, March 1991, 18—22.

7. Domenico Ferrari, Giuseppe Serazzi, and Alessandro Zeigner, *Measurement and Tuning of Computer Systems*, Prentice-Hall, Englewood Cliffs, NJ, 1983.

8. Martin Reiser, "Mean value analysis of queueing networks, A new look at an old problem," *Proc. 4th Int. Symp. on Modeling and Performance Evaluation of Computer Systems*, Vienna, 1979.

9. Martin Reiser, "Mean value analysis and convolution method for queue-dependent servers in closed queueing networks," *Performance Evaluation*, 1(1), January 1981, 7—18

10. Martin Reiser and Stephen S. Lavenberg, "Mean value analysis of closed multichain queueing networks," *JACM*, 22, April 1980, 313—322.

Chapter 5 Model Parameterization

The wind and the waves are always on the side of the ablest navigators.
Edward Gibbon

You know my methods, Watson.
Sherlock Holmes

5.1 Introduction

In this chapter we examine the measurement problem and the problem of parameterization. The measurement problem is, "How can I measure how well my computer system is processing the workload?" We assume that you have one or more measurement tools available for your computer system or systems. We discuss how to use your measurement tools to find out what your computer system is doing from a performance point of view. We also discuss how to get the data you need for parameterizing a model. In many cases it is necessary to process the measurement data to obtain the parameters needed for modeling.

5.2 Measurement Tools

The basic measurement tool for computer performance is the monitor. There are two basic types of monitors: software monitors and hardware monitors. Hardware monitors are used almost exclusively by computer manufacturers.

Hardware monitors are electronic devices that are connected to computer systems by probes attached to points in the system such as busses and registers. They operate by sensing and recording electrical signals. Ferrari *et al.* in Section 5.3 of [Ferrari, Serazzi, and Zeigner 1983] discuss some applications of hardware monitors such as the measurement of the seek activity of a disk unit. The main advantages of a hardware monitor over a software monitor are (1) no overhead on the resources of the computer system such as CPU or memory, (2) better time resolution since hardware monitors have internal clocks with resolutions in the nanosecond range while software monitors usually use a system clock with millisecond resolutions, and (3) higher sampling rates (we discuss sampling later). The overwhelming disadvantage for most installations is the high cost and the need for special expertise to use a hardware monitor effectively. Most readers of this book will not be concerned with hardware monitors.

There are other detailed classifications of performance monitors but we restrict our discussion to software monitors because they are the concern of almost all performance managers. The three most common types of software monitors are used for diagnostics (sometimes called real-time or trouble shooting monitors), for studying long-term trends (sometimes called historical monitors), and job accounting monitors for gathering chargeback information. These three types can be used for monitoring the whole computer system or be specialized for a particular piece of software such as CICS, IMS, or DB2 on an IBM mainframe. There are probably more specialized monitors designed for CICS than for any other software system.

The uses for a diagnostic monitor include the following:

1. To determine the cause of poor performance at this instant.
2. To identify the user(s) and/or job(s) that are monopolizing system resources.
3. To determine why a batch job is taking an excessively long time to complete.
4. To determine whether there is a problem with the database locks.
5. To help with tuning the system.

To accomplish these uses a diagnostic monitor should first present you with an overall picture of what is happening on your system plus the ability to focus on critical areas in more detail. A good diagnostic monitor will provide assistance to the user in deciding what is important. For example, the monitor may highlight the names of jobs or processes that are performing poorly or that are causing overall systems problems. Some diagnostic monitors have expert system capabilities to analyze the system and make recommendations to the user.

A diagnostic monitor with a built-in expert system can be especially useful for an installation with no resident performance expert. An expert system or adviser can diagnose performance problems and make recommendations to the user. For example, the expert system might recommend that the priority of some jobs be changed, that the I/O load be balanced, that more main memory or a faster CPU is needed, etc. The expert system could reassure the user in some cases as well. For example, if the CPU is running at 100% utilization but all the interactive jobs have satisfactory response times and low priority batch jobs are running to fully utilize the CPU, this could be reported to the user by the expert system.

Uses for monitors designed for long-term performance management include the following:

1. To archive performance data for a performance database.
2. To provide performance information needed for parameterizing models of the system.
3. To provide performance data for forecasting studies.

Most of the early performance monitors were designed to provide information for chargeback. One of the most prominent of these is the System Management Facility discussed by Merrill in [Merrill 1984] as follows:

> System Management Facility (SMF) is an integral part of the IBM OS/360, OS/VS1, OS/VS2, MVS/370, and MVS/XA operating systems. Originally called System Measurement Facility, SMF was created as a result of the need for computer system accounting caused by OS/360. A committee of the SHARE attendees and IBM employees specified the requirements, which were then implemented by IBM and were generally available with Release 18 of OS/360. The SHARE Computer Management and Evaluation Project is the direct descendant of this original 1969 SHARE committee.

As Merrill points out, SMF information is also used for computer performance evaluation.

Accounting monitors, such as SMF, generate records at the termination of batch jobs or interactive sessions indicating the system resources consumed by the job or session. Items such as CPU seconds, I/O operations, memory residence time, etc., are recorded.

Two software monitors produced by the Hewlett-Packard Performance Technology Center are used to measure the performance of the HP-UX system I am using to write this book. HP GlancePlus/UX is an online diagnostic tool (sometimes called a trouble shooting tool) that monitors ongoing system activity. The HP GlancePlus/UX User's Manual provides a number of examples of how this monitor can be used to perform diagnostics, that is, determine the cause of a performance problem. The other software monitor used on the system is HP LaserRX/UX. This monitor is used to look into overall system behavior on an ongoing basis, that is, for trend analysis. This is important for capacity planning. It is also the tool we use to provide the information needed to parameterize a model of the system.

There are two parts of every software monitor, the collector that gathers the performance data and the presentation tools designed to present the data in a meaningful way. The presentation tools usually process the raw data to put it into a convenient form for presentation. Most early monitors were run as batch jobs and the presentation was in the form of a report, which also was generated by a batch job. While monitor collectors for long range monitors are batch jobs, most diagnostic monitors collect performance data only while the monitor is activated.

The two basic modes of operation of software monitors are called *event-driven* and *sampling*. Events indicate the start or the end of a period of activity or inactivity of a hardware or software component. For example, an event could be the beginning or end of an I/O operation, the beginning or end of a CPU burst of activity, etc. An event-driven monitor operates by detecting events. A sampling monitor operates by testing the states of a system at predetermined time intervals, such as every 10 ms. A sampling monitor would find the CPU utilization by checking the CPU every t seconds to find out if it is busy or not. Clearly, the value of t must be fairly small to ensure the accuracy of the measurement of CPU utilization; it is usually on the order of 10 to 15 milliseconds. A small value of t means sampling occurs fairly often, which increases sampling overhead. CPU sampling overhead is typically in the range of 1 to 5 percent, that is, the CPU is used 1 to 5 percent of the time to perform the sampling. Ferrari *et al.* in Chapter 5 of [Ferrari, Serazzi, and Zeigner 1983] provide more details about sampling overhead.

Software monitors are very complex programs that require an intimate knowledge of both the hardware and operating system of the computer system being measured. Therefore, a software monitor is usually purchased from the computer company that produced the computer being monitored or a software performance vendor such as Candle Corporation, Boole & Babbage, Legent, Computer Associates, etc. For more detailed information on available monitors see [Howard Volume 2].

If you are buying a software monitor for obtaining the performance parameters you need for modeling your system, the properties you should look for include:

1. Low overhead.

2. The ability to measure throughput, service times, and utilization for the major servers.

3. The ability to separate workload into homogeneous classes with demand levels and response times for each.
4. The ability to report metrics for different types of classes such as interactive, batch, and transaction.
5. The ability to capture all activity on the system including system overhead by the operating system.
6. Proviside sufficient detail to detect anomalous behavior (such as a runaway process), which indicates atypical activity.
7. Provide for long-term trending via low volume data.
8. Good documentation and training provided by the vendor.
9. Good tools for presenting and interpreting the measurement results.

Low overhead is important both because it leaves more capacity available for performing useful work and because high overhead distorts the measurements made by the monitor.

The problem of measuring system CPU overhead has always been a challenge at IBM MVS installations. It is often handled by "capture ratios." The capture ratio of a job is the percentage of the total CPU time for a job that has been captured by SMF and assigned to the job. The total CPU time consists of the TCB (task control block) time plus the SRB (system control block) time plus the overhead, which normally cannot be measured. It may require some less than straightforward calculations to convert the measured values of TCB and SRB provided by SMF records into actual times in seconds. For an example of these calculations see [Bronner 1983]. For an overview of RMF see [IBM 1991]. If the capture ratio for a job or workload class is known, the total CPU utilization can be obtained by dividing the sum of the TCB time and the SRB time by the capture ratio. The CPU capture ratio can be estimated by linear regression and other techniques. Wicks describes how to use the regression technique in Appendix D of [Wicks 1991]. The approximate values of the capture ratio for many types of applications are known. For example, for CICS it is usually between 0.85 and 0.9, for TSO between 0.35 and 0.45, for commercial batch workload classes between 0.55 and 0.65, and for scientific batch workload classes between 0.8 and 0.9.

Example 5.1

The performance analysts at Black Bart measure their MVS system over a period of 4,500 seconds with RMF and find that the measured total CPU time is 2,925 seconds so the average CPU utilization over the period is 2,925/4,500 = 0.65 or 65 percent. However, the total CPU time reported for the two workload classes, wk1 and wk2, is 1,800 seconds and 675 seconds, respectively. Since these numbers add up to 2,475 seconds, 450 seconds are not accounted for and thus must be assumed to be overhead. If the analysts do not know the capture ratios for the two workload classes, the usual procedure is to assign the overhead proportionally, that is, assign ((1,800/(1,800 + 675))(450) = 316 seconds to wk1 and the other 134 seconds to wk2. Then, over the 4,500-second interval wk1 has (1,800 + 316)/4,500 = 0.47 or 47 percent CPU utilization and wk2 has (675 + 134)/4,500 = 0.18 or 18 percent CPU utilization. This means the effective capture ratio for each class is 0.55/0.65. On the other hand, if the Black Bart performance analysts had previously found that the capture ratio for wk1 was approximately 0.9 and for wk2 it was 0.85, then they would assign 1,800/0.9 = 2,000 CPU seconds to wk1 and 675/0.85 = 794 seconds to wk2 even though the sum is not exactly 2,925 seconds. According to Bronner [Bronner 1983], if the sum of all the CPU times estimated from the use of capture ratios is within 10 percent of the actual CPU utilization, the CPU estimates are acceptable. Here the error is only 4.48 percent.

Monitors are able to accumulate huge amounts of data. It is important to have facilities for reducing and presenting this data in an understandable format. One of the most common ways of presenting information, such as global CPU utilization, is by means of graphs showing the evolution of the measurement(s) over time. In Figure 5.1 we can see parts of a couple of graphs and a display table from a software monitor. The table shows that at 11 am on April 3, 1991, the application called "system notes" was consuming 17.5 percent of the CPU on the HP-UX system being monitored by HP LaserRX/UX. The reason for displaying the very detailed table was that the graph above it indicated that the Global System CPU Utilization was very high at 11 am on April 3. The use of this graph in turn was triggered by the study of the Global Bottlenecks graph. Thus in using monitors one normally proceeds from the general to the specific.

Figure 5.1. Monitor Presentation of Example 5.1

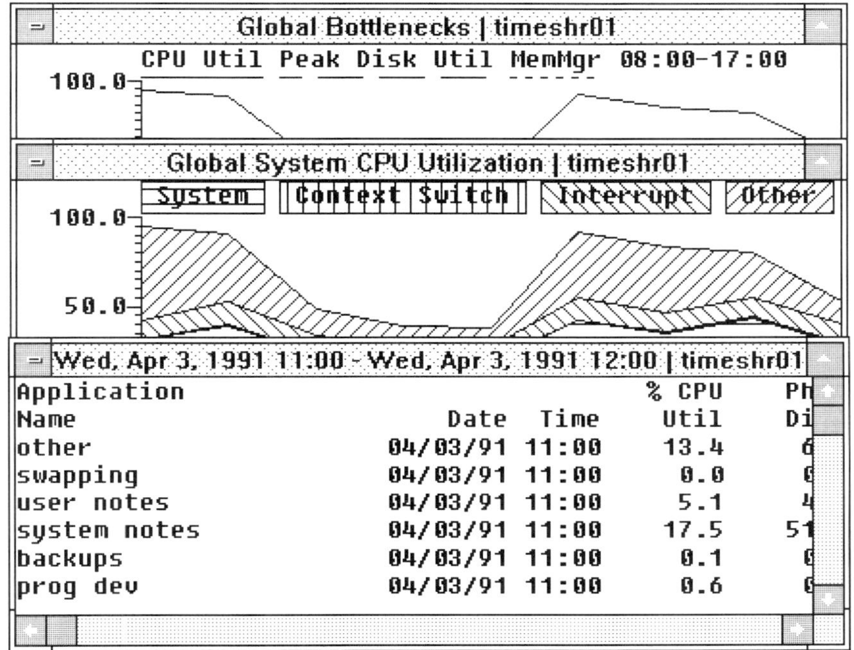

When you arrive at a fork in the road, take it.
Yogi Berra

5.3 Model Parameterization

Model parameterization is an important part of any modeling study. The accuracy of the results depends upon the accuracy of the parameter values. In addition, modeling studies are carried out by modifying parameter values to project the performance of modified systems.

While modeling of proposed new systems by computer manufacturers is an important part of modeling, we restrict our discussion to that of studying an existing system. We assume that the purpose of the modeling study is to investigate the effect on performance of an existing system due to changes in the configuration or workload.

5.3.1 The Modeling Study Paradigm

We discussed the general modeling study paradigm in Section 3.5 of Chapter 3. We will examine it in more detail here.

A modeling study of an existing system consists of the following steps:

1. Define the purpose of the modeling study.
2. Decide what period of the day to measure and model.
3. Make measurements of the current system to determine the performance and to obtain the parameters for the model.
4. Parameterize the model and use the model to predict the current performance.
5. Compare the predicted current performance with the measured performance and adjust the model until there is satisfactory agreement.
6. Modify the inputs to the model to make performance predictions for the modified system.
7. After the system is modified compare the measured performance with the predicted performance.

Although Steps 1 and 7 are very important, these steps tend to be the most neglected.

Failure to specify carefully the purpose of a modeling study is an almost surefire guarantee of failure. The purpose of the study colors the measurements taken, the method of analysis, the assumptions made, the resources used, the reports to management, and other considerations too numerous to catalog.

An example of the purpose of a modeling study is: "Can the workloads running on two separate Hewlett-Packard HP 3000 Series 980/100 uniprocessors be combined to run on one HP 3000 Series 980/300 multiprocessor?" The Series 980/300 has three processors and is rated as roughly 2.1 times as powerful as a Series 980/100.To answer this question, the hardware and software of the three computers in question must be completely specified, the workloads carefully defined, and the performance criteria for measuring whether or not the combined workload can run on one Series 980/300 must be chosen.

Step 7 in the modeling paradigm is an opportunity to learn from the study. If the predicted performance of the modified system is quite different from the actual measured performance, it is important to find out why. Often the difference is due to errors in predicting the load on the modified system. For example, it might have been necessary to schedule work on the modified system that had not been anticipated. It may have been due to modeling errors. If this is true, it is important to correct the errors so that future modeling studies can be improved.

For Step 2 we must decide what measurement period to use for the model. Analysts usually choose a peak period of the day, week, or month since this is when performance problems are most likely to exist. The length of the measurement interval is also very important because of the problem of *end effects*. End effects are measurement errors caused because some of the customers are processed partly outside the measurement interval. Longer intervals have less error from end effects than shorter intervals. Intervals from 30 to 90 minutes are typical intervals chosen because they are long enough to keep end effects under control and short enough to keep the amount of data needed in reasonable balance.

5.3.2 Calculating the Parameters

The first step in determining the parameters for a model is to determine what workload classes are to be used and what type. Recall that, from Chapter 3, the three types of workload classes are transaction, batch, and terminal. We assume that C is the number of workload classes. Each workload class is characterized by its workload intensity and by the service demands at each of the K service centers of the model. For each class c its workload intensity is one of:

> λ_c, the average arrival rate (for transaction workloads), or
> N_c, the population (for batch workloads), or
> N_c and Z_c, the number of terminals and the think time (for terminal workloads).

For each work load the service demand for each class c and center k is $D_{c,k}$, the service demand or total service time required at center k by workload c.

Some modeling software has the capability of automating the parameterization of the model. However, the person running the modeling package must still get involved in the validation process ,which can lead to changes in the modeling setup. Two modeling packages that have the automated modeling capability are Best/1 MVS from BGS Systems and MAP from Amdahl Corporation. Both model IBM mainframes running the MVS operating system.

Best/1 MVS uses the CAPTURE/MVS data reduction and analysis tool. By combining data from two standard measurement facilities (RMF and SMF), CAPTURE/MVS reports contain both system-wide use of hardware resources and workload specific performance measures. In addition, CAPTURE/MVS also automatically produces input to BEST/1 MVS. By using the AUTO-CAPTURE facility, new or infrequent users need not learn the command syntax and associated JCL statements and thus save a lot of time and effort.

For MAP users the automated method uses the OBTAIN feature of MAP. This facility, available only for MVS installations, allows SMF/RMF data to be processed and a MAP model generated. OBTAIN processes the SMF/RMF data and constructs a system model based on both the information contained in these records, and on user-provided parameters that specify how workload data in SMF records is to be interpreted. The OBTAIN feature is a separate application program within the MAP product that executes interactively. Stoesz [Stoesz 1985] discusses the validation process after using CAPTURE/MVS or OBTAIN to construct an analytical queueing model of an MVS system.

In the following example we assume that the performance information available is similar to that provided by SMF and RMF records on an IBM mainframe running under the MVS operating system. We have used the technical bulletins [Bronner 1983] and [Wicks 1991] as guides for this example. We assume that for terminal workload classes the average number of active terminals, the average number of interactions completed, the average response time, and the average service demand of the workload class for each service center is provided or can be obtained without excessive calculation. Then, from the number of interactions completed in the observation interval, we calculate the average throughput $X_c = \lambda_c$. (This is an approximation due to end effects.) We estimate the average think time from the response time formula as follows:

$$Z_c = \frac{N_c}{X_c} - R_c.$$

For batch workload classes we assume we are provided with the average number of jobs in service, the number of completions, the average turnaround time, and all service demands.

Example 5.2

A small computer system at Big Bucks Bank was measured using their software performance monitor for 1 hour and 15 minutes (4,500 seconds). The computer system has three workload classes, two terminal and one batch. The terminal classes are numbered 1 and 2 with the batch class assigned number 3. Some of the measurement results are shown in Tables 5.1 through 5.3.

Table 5.1. Example 5.2 Data

c	N_c	Interactions	R_c
1	10.1	1485	0.20
2	4.9	1062	1.15
3	2.2	6570	1.41

They also obtained the device utilization and average number of customers at each of the three devices as shown in Table 5.2. The CPU utilization has been corrected for any capture ratio errors, that is, the CPU utilization accounts for CPU overhead.

Table 5.2. More Example 5.2 Data

k	Number	Utilization
1	2.06	0.93
2	0.16	0.13
3	0.22	0.18

Table 5.3 provides the measured service demands for each job class at the CPU and each of the two I/O devices.

Table 5.3. Still More Example 5.2 Data

c	k	$D_{c,k}$
1	CPU	0.025
1	I/O 1	0.040
1	I/O 2	0.060
2	CPU	0.200
2	I/O 1	0.200
2	I/O 2	0.060
3	CPU	0.600
3	I/O 1	0.050
3	I/O 2	0.060

We show the preliminary calculations the performance analysts at Big Bucks Bank made with *Mathematica* to prepare for modeling the baseline system. Note that the throughput of each class is calculated by dividing the number of completed interactions or jobs by the time; in this case the time is 4,500 seconds. The throughput formula is then used to calculate the mean think time. Then we show how the program **Approx** is used to calculate the performance numbers from the measured data. This is part of the initial validation procedure.

```
In[4]:= x1 = 1485./4500

Out[4]= 0.33

In[5]:= z1 = 10.1/x1 -0.2

Out[5]= 30.4061

In[6]:= x2 = 1062./4500
Out[6]= 0.236
```

```
In[7]:= z2 = 4.9/x2 - 1.15

Out[7]= 19.6127

In[8]:= x3 = 6570./4500

Out[8]= 1.46

In[9]:= n3 = 1.46/ x3

Out[9]= 1.

In[10]:= n3 = 1.41 x3

Out[10]= 2.0586

In[11]:= Pop = {10.1, 4.9, n3}

Out[11]= {10.1, 4.9, 2.0586}

In[12]:= Think = {z1, z2, 0}

Out[12]= {30.4061, 19.6127, 0}

In[13]:= Demands = {{.025, .04, .06},{.2, .2, .06},
{.6, .05, .06}}

Out[13]= {{0.025, 0.04, 0.06}, {0.2, 0.2, 0.06}, {0.6,
0.05, 0.06}}

In[14]:= Demands[[2]] ={.2, .2, .3}

Out[14]= {0.2, 0.2, 0.3}

In[15]:= Demands

Out[15]= {{0.025, 0.04, 0.06}, {0.2, 0.2, 0.3}, {0.6,
0.05, 0.06}}
```

```
In[16]:= Approx[Pop, Think, Demands, 0.001]

Class#   Think    Pop            Resp          TPut
------   -----   -----   -------------   -----------
  1     30.4061   10.1       0.194436       0.33006
  2     19.6127    4.9       1.188481      0.235563
  3        0.0   2.0586      1.403805      1.466443

Center#             Number Utilization
 -------------------- -----------
        1          2.042018     0.93523
        2          0.150296    0.133637
        3          0.210424    0.178459
```

The analysts at Big Bank feel that the model outputs are sufficiently close to the measured values to validate the model. They are satisfied with the current performance of the computer system but the users have told them that the throughput of the first online system will quadruple and the throughput of the second online workload will double in the next six months, although the batch component is not expected to increase. The analysts feel that an upgrade to a computer with a CPU that is 1.5 times as fast without changing the I/O might satisfy the requirements of their users. The users want to be able to process the new volume of online work without increasing the response time of the first workload class above 0.2 seconds and that of the second workload class above 1.0 seconds with the turnaround time of the batch workload remaining below 1.0 seconds. The analysts model the proposed system using the *Mathematica* program **Fixed** as follows:

```
In[4]:= Demands = {{.025/1.5, .04, .06}, {.2/1.5, .2,
.3}, {.6/1.5, .05, .06}}

Out[4]= {{0.0166667, 0.04, 0.06}, {0.133333, 0.2,
0.3}, {0.4, 0.05, 0.06}}

In[5]:= Think = {30.4061, 19.6127, 0}

Out[5]= {30.4061, 19.6127, 0}

In[6]:= x1=4 .33

Out[6]= 1.32
```

```
In[7]:= x2 = 2 0.236

Out[7]= 0.472

In[8]:= x3 = 1.46

Out[8]= 1.46

In[9]:= Fixed[{x1, x2, x3}, {,,}, Think, Demands,
0.001]


  Class#          ArrivR            Pc
-------- -------------- ------------
       1           1.32      40.3563
       2          0.472      9.68986
       3           1.46     0.876127


  Class#            Resp            TPut
-------- -------------- -------------
       1        0.16682           1.32
       2       0.916657          0.472

  Center#         Number    Utiliz
--------- ------------- ---------
        1      0.829247  0.668933
        2      0.272689    0.2202
        3      0.427057    0.3084
```

Note that the response time requirements are far exceeded. Perhaps Big Bucks could make do with a slightly smaller processor. Note, also, that there will be approximately 40.3563 active users of the first online application, 9.68986 active users of the second online application, and 0.876127 active batch jobs with the new system.

Exercise 5.1

Ross Ringer, a fledgling performance analyst at Big Bucks Bank, suggests that they could save a lot of money by procuring the model of their current machine with a CPU 25 percent faster than their current machine rather than one that is 50 percent faster. This machine could then be "board upgraded" to a CPU with twice the power of the current machine for a very reasonable price. By "board upgraded" we mean that the old CPU board could be replaced with the faster CPU board without changing any of the other components. Use **Fixed** to see if Ross is right.

Exercise 5.2

Fruitful Farms measures the performance of one of their computer systems during the peak afternoon period of the day for 1 hour (3,600 seconds). Their monitor reports that the CPU is idle for 600 seconds of this interval and thus busy for 3,000 seconds (50 minutes). Fruitful Farms has three workload classes on the computer system, one terminal class, term, and two batch classes, batch1 and batch2. The monitor reports that workload class term used 20 minutes of CPU time, batch1 used 8 minutes, and batch2 used 2 minutes. (a) Calculate the amount of the 3000 seconds of CPU time that should be allocated to each workload class assuming the capture ratio is the same for all workloads. (b) Make the calculation of part (a) assuming that all CPU overhead is due to paging and that 80% of the paging is for the terminal class while 15% is for batch1 and 5% for batch2.

5.4 Solutions

Solution to Exercise 5.1

Ross calculates the new service demands for the CPU for the three workload classes by multiplying each of the demands for the upgraded CPU in Example 5.2 by 1.5/1.25, yielding the values shown in the matrix Demands displayed in the following *Mathematica* session:

```
In[19]:= MatrixForm[Demands]

Out[23]//MatrixForm= 0.02   0.04   0.06

                     0.16   0.2    0.3

                     0.48   0.05   0.06

In[24]:= Think

Out[24]= {30.4061, 19.6127, 0}
```

```
In[25]:= x = {x1, x2, x3}

Out[25]= {1.32, 0.472, 1.46}

In[26]:= Fixed[x, {,,}, Think, Demands, 0.001]

  Class#         ArrivR            Pc
-------- -------------- ------------
       1           1.32      40.3729
       2          0.472      9.73681
       3           1.46      1.13553


  Class#           Resp          TPut
-------- -------------- -------------
       1       0.179405          1.32
       2       1.016144         0.472
       3       0.777757          1.46


  Center#        Number    Utiliz
--------- ------------- ---------
        1      1.149905   0.80272
        2      0.273591    0.2202
        3      0.428464    0.3084
```

From the output above we see that Ross is almost right! The average response time for the second online workload class is 1.016144 seconds, which is slightly over the 1.0-second goal. However, this is an approximate model and all the estimates are approximate as well, so Ross's recommendation is OK.

Solution to Exercise 5.2

For part (a) we note that the reported fraction of CPU used by the three classes is 20/30, 8/30, and 2/30, respectively. The unallocated CPU time of 1,200 seconds should be allocated in the same ratio. Hence, as shown in the following *Mathematica* calculations , we allocate 800 seconds, 320 seconds, and 80 seconds, respectively, to the three classes. This means the total CPU times for the three classes are 33 minutes and 20 seconds; 13 minutes and 20 seconds; and 3 minutes and 20 seconds.

```
In[45]:= 1200 20/30

Out[45]= 800
In[46]:= 1200 8/30

Out[46]= 320

In[47]:= 1200 2/30

Out[47]= 80

In[48]:= (20 60 + 800)/60

         100
Out[48]= ---
          3

In[49]:= N[%]

Out[49]= 33.3333

In[50]:= (8 60 + 320)/60

         40
Out[50]= --
         3

In[51]:= N[%]

Out[51]= 13.3333

In[52]:= (2 60 + 80)/60

         10
Out[52]= --
         3
```

For part (b) we allocate 80% of the 1,200 unallocated CPU seconds to the term workload class; this comes to 960 seconds or 16 minutes. We allocate 15% of 1200 or 180 seconds (3 minutes) to batch1 and the other 5% or 1 minute to batch2. The *Mathematica* calculations for this follow:

```
In[55]:= .8 1200

Out[55]= 960.
In[56]:= %/60

Out[56]= 16.

In[57]:= .15 1200

Out[57]= 180.

In[58]:= %/60

Out[58]= 3.
```

5.5 References

1. Leroy Bronner, *Capacity Planning: Basic Hand Analysis*, IBM Washington Systems Center Technical Bulletin, December 1983.

2. Domenico Ferrari, Giuseppe Serazzi, and Alessandro Zeigner, *Measurement and Tuning of Computer Systems*, Prentice-Hall, Englewood Cliffs, NJ, 1983.

3. Phillip C. Howard, *IS Capacity Management Handbook Series, Volume 1, Capacity Planning,* Institute for Computer Capacity Management, updated every few months.

4. Phillip C. Howard, *IS Capacity Management Handbook Series, Volume 2, Performance Analysis and Tuning,* Institute for Computer Capacity Management, updated every few months.

5. IBM, *MVS/ESA Resource Measurement Facility Version 4 General Information,* GC28-1028-3, IBM, March 1991.

6. H. W. "Barry" Merrill, *Merrill's Expanded Guide to Computer Performance Evaluation Using the SAS System*, SAS, Cary, NC, 1984.

7. Roger D. Stoesz, "Validation tips for analytic models of MVS systems," *CMG '85 Conference Proceedings,* Computer Measurement Group, 1985, 670—674.

8. Raymond J. Wicks, *Balanced Systems and Capacity Planning,* IBM Washington Systems Center Technical Bulletin GG22-9299-03, September 1991.

Chapter 6 Simulation and Benchmarking

Monte Carlo Method *[Origin: after Count Montgomery de Carlo, Italian gambler and random-number generator (1792–1838).] A method of jazzing up the action in certain statistical and number-analytic environments by setting up a book and inviting bets on the outcome of a computation.*

Stan Kelly-Bootle
The Devil's DP Dictionary

Benchmark v.trans *To subject (a system) to a series of tests in order to obtain prearranged results not available on competitive systems. See also MENDACITY SEQUENCE.*

Stan Kelly-Bootle
The Devil's DP Dictionary

The purpose of computing is insight, not numbers.
Richard V. Hamming

6.1 Introduction

Simulation and benchmarking have a great deal in common. When simulating a computer system we manipulate a model of the system; when benchmarking a computer system we manipulate the computer system itself. Manipulating the real computer system is more difficult and much less flexible than manipulating a simulation model. In the first place, we must have physical possession of the computer system we are benchmarking. This usually means it cannot be doing any other work while we are conducting our benchmarking studies. If we find that a more powerful system is needed we must obtain access to the more powerful system before we can conduct benchmarking studies on it. By contrast, if we are dealing with a simulation model, in many cases, all we need to do to change the model is to change some of the parameters.

For benchmarking an online system, in most cases, part of the benchmarking process is simulating the online input used to drive the benchmarked system. This is called "remote terminal emulation" and usually is performed on a second computer system which transmits the simulated online workload to the computer under study. In some cases the remote terminal emulation is performed on the machine that is being benchmarked but this creates special problems in evaluating the benchmark. The simulator that performs the remote terminal emulation is called a *driver.* The most representative online benchmarking is achieved by having real people key in the workload in the form of scripts as the benchmark is run; this is prohibitively expensive in most cases. In addition, a benchmark session of this type is not repeatable; a person cannot key in a script twice in exactly the same way. For these reasons remote terminal emulation is the method most commonly used to simulate the online workload classes. Thus simulation modeling is also part of benchmark modeling for most benchmarks that include terminal workloads.

Another common feature of simulation and benchmarking is that a simulation run and a benchmarking run are both examples of a random process and thus must be analyzed using statistical analysis tools. The proper analysis of simulation output and benchmarking output is a key part of simulation or benchmarking; such a study without proper analysis can lead to the wrong conclusions.

Simulation is better than reality!
Richard W. Hamming

6.2 Introduction to Simulation

There are a number of kinds of simulation including Monte Carlo simulation, the kind of simulation described in the quote at the beginning of the chapter. Monte Carlo simulation is used to solve difficult mathematical problems not amenable to analytic solution. While some simulation experts restrict the name "Monte Carlo simulation" to this type of simulation, Knuth in his widely referenced book [Knuth 1981] says, "These traditional uses of random numbers have suggested the name 'Monte Carlo method,' a general term used to describe any algorithm that employs random numbers." The kind of simulation that is most important for modeling computer systems is often called *discrete event simulation* but certainly falls within the rubric of what Knuth calls the Monte Carlo method.

Simulation is a very powerful modeling technique. It is used to build flight trainers for budding flyers as well as for training experienced pilots on planes; to study theories in physics, cosmology, and other disciplines; and to model computer systems. After the crash of a DC-10 aircraft near Chicago a few years ago because an engine fell off, a DC-10 flight training simulator was used to study whether or not the plane could be controlled with one engine detached. (It could but the pilots did not realize they had lost an engine until too late.) For other exotic applications of simulation see [Pool 1992].

Twenty years ago modeling computer systems was almost synonymous with simulation. Since that time so much progress has been made in analytic queueing theory models of computer systems that simulation has been displaced by queueing theory as the modeling technique of choice; simulation is now considered by many computer performance analysts to be the modeling technique of last resort. Most modelers use analytic queueing theory if possible and simulation only if it is very difficult or impossible to use queueing theory. Most current computer system modeling packages use queueing network models that are solved analytically. Some of the best known of these are Best/1 MVS from BGS Systems, Inc.; MAP from Amdahl Corp.; CA-ISS/THREE from Computer Associates, International, Inc.; and Model 300 from Boole & Babbage. RESQ from IBM provides both simulation and analytic queueing theory modeling capabilities.

The reason for the preference by most analysts for analytic queueing theory modeling is that it is much easier to formulate the model and takes much less computer time to use than simulation. See, for example, the paper [Calaway 1991] we discussed in Chapter 1. Kobayashi in his well-known book [Kobayashi 1978] says:

> It is quite often found, however, that a simulation model takes much longer to construct, requires much more computer time to execute, and yet provides much less information than the model writer expected. Therefore, simulation should generally be considered a technique of last resort. Yet, many problems associated with design and configuration changes of computing systems are so complex that an analytical approach is often unable to characterize the real system in a form amenable to solution. Consequently, despite its difficulties and the costs and time required, simulation is often the only practical solution to a real problem.

To perform steps 4 and 5 of the modeling study paradigm described in Section 5.3.1 (and more briefly in Section 3.5) requires the following basic tasks.

1. Construct the model by choosing the service centers, the service center service time distributions, and the interconnection of the center.

2. Generate the transactions (customers) and route them through the model to represent the system.

3. Keep track of how long each transaction spends at each service center. The service time distribution is used to generate these times.

4. Construct the performance statistics from the above counts.

5. Analyze the statistics.

6. Validate the model.

Of course, these same tasks are necessary for Step 6 of the modeling study paradigm.

One of the major activities in any simulation study is writing the computer code that makes the calculations for the study. Such programs are called *simulators*. In the next section we discuss how simulators are written.

6.3 Writing a Simulator

As we mentioned in the last section, a simulator is a computer program written to construct a simulation model. One of the best references on simulator design is the chapter *Simulator Design and Programming* by Markowitz in [Lavenberg 1983]. Markowitz is not only the developer of the first version of SIMSCRIPT, an early simulation language, but also has a Nobel prize in economics!

To illustrate the challenges of simulation let us consider the *Mathematica* program **simmm1** for simulating an M/M/1 queueing system. The M/M/1 queueing system is the simplest queueing system that is in widespread use. Kleinrock in his classic book [Kleinrock 1975] refers to the M/M/1 queueing system as follows:

> ...the celebrated M/M/1 queue is the simplest nontrivial interesting system and may be described by selecting the birth-and-death coefficients as follows:

The M/M/1 queueing system is an open system with one server that provides exponentially distributed service; this means that the probability that the provided service will require not more than t time units is given by $P[s \le t] = 1 - e^{-t/S}$

where S is the average service time. For the M/M/1 queueing system the interarrival time, that is, the time between successive arrivals, also has an exponential distribution. Thus, if τ describes the interarrival time, then $P[\tau \le t] = 1 - e^{-\lambda t}$, where λ is the average arrival rate. The two parameters that define this model are the average arrival rate (customers per second) λ, and the average service time S (seconds per customer).

```
simmm1[lambda_Real, serv_Real, seed_Integer, n_Inte-
ger, m_Integer]:=
Block[{t1, t2, s, s2, t, i, j, k, lower, upper, v, w,
h},
SeedRandom[seed];
t1=0;
t2=0;
s2=0;
For[w=0; i = 1, i<=n, i++,
s = - serv Log[Random[]];
t = - (1/ lambda) Log[Random[]];
If[w<t, w = s, w = w + s -t];
s2 = s2 + w];
Print["The average value of response time at end of
warmup is ", N[s2/n, 5]];
t1=0;
t2=0;
For[j=1, j<=100, j++,
s2=0;
For[k=1, k<=m, k++,
t = - (1/lambda) Log[Random[]];
s = - serv Log[Random[]];
If[w<t, w =s, w = w + s - t];
s2 = s2 + w];
t1 = t1 + s2/m;
t2 = t2 + (s2/m)^2];
v = (t2 - (t1^2)/100)/99;
h = 1.984217 Sqrt[v]/10;
lower = t1/100 - h;
upper = t1/100 + h;
Print["Mean time in system is ",N[t1/100, 6]];
Print["95 percent confidence interval is"];
Print[lower, " to " ,upper];
]
```

One of the problems with simulation is determining when the simulation process has reached the steady-state. When a simulator is executed by moving customers through it, the outputs (queue lengths, utilizations, and subsystem response times) go through a *transient phase*, which depends upon the initial conditions, and finally reaches a limiting *steady-state* or *equilibrium* condition in which the distributions of the outputs are independent of the initial conditions. By "initial conditions" we mean the number of customers at each service center at the beginning of a simulation run. Usually, simulators use the initial condition that all queues and service centers are empty. Of course, other choices usually have to be made as well to define the initial conditions. If you have trouble with the concept of steady-state, do not despair. It is a very sophisticated concept. The best explanation that I've seen is given by Welch [Welch 1983]; Welch provides some very helpful graphics to illustrate what happens during a simulation. The information from the transient part of the simulation is usually ignored in calculating the outputs from the simulation study. No one has been able to find a general rule or procedure that will always guarantee that the steady-state has been reached, although Kobayashi [Kobayashi 1978] has developed some rules for some special cases. MacDougall [MacDougall 1987] makes some recommendations for the length of a warmup run, that is, the first part of the simulation that gets the system into the steady-state. In **simmm1**, we assume that the M/M/1 queueing system has reached the steady-state when n customers have been served and leave it to the user to choose the value of n. We begin to compile our statistics at this point; that is, we ignore the statistics for the first n customers.

Bookkeeping is another special problem for writing a simulator. By bookkeeping we mean keeping track of how much time each customer spends in each service facility as well as scheduling the beginning and end of each service. Even for this simple M/M/1 system, keeping track of the time spent in the system for each customer requires some care.

Generating random sequences of specified types is also very much a part of constructing a simulator. For **simmm1** we generated two random sequences of exponentially distributed random numbers, one for the interarrival times and one for the service times. To generate a sequence of random numbers with an exponential distribution with average value 10, for example, using *Mathematica* we need only repeatedly use the statement $s = -10$ Log[Random[]]. Therefore, we could generate 20 such numbers as follows:

```
In[3]:= Table[-10 Log[Random[]], {20}]

Out[3]= {4.00606, 15.0269, 4.21232, 5.31992, 1.08033,
```

```
10.5912, 6.6391,

>     17.1118, 0.80239, 28.088, 0.666785, 3.89245,
3.85219, 19.1179, 13.3461,

>     40.1615, 3.78502, 18.3989, 8.93976, 3.00079}

In[4]:= Apply[Plus, %]/20
Out[4]= 10.402
```

The **Random** function in *Mathematica* chooses a random number that is between zero and one. **Random** depends upon a starting value of an algorithm; this starting value is called the *seed*. If we want to make different runs of **simmm1** yield different results, we change the seed; if we want to repeat a run exactly we use the same seed.

In **simmm1** we use the *method of batch means* to calculate not only an estimated average response time for the system, which we call the *mean response time* in the code, because mean and average mean the same thing, but also a 95 percent confidence interval for the average value. The idea of the method of batch means is to first make a warmup run to put the simulation process into the steady-state (some authorities leave out the warmup run) followed by several runs in sequence. In each of the runs the average values of important parameters are estimated. Then, by comparing the averages estimated in the different runs, a confidence interval for each can be calculated. In **simmm1** we have set it up so that 100 independent runs are made after the warmup run. From these 100 runs a 95 percent confidence interval for the average response time is calculated. A 95 percent confidence interval for the average response time (mean response time) is an interval such that, if a large number of simulation runs similar to the current run are made, then 95 percent of the time the true steady-state average value (mean) will be inside the interval and 5 percent of the time it won't. A short confidence interval means that we can be more confident that our result is close to the exact value than we would have for a long confidence interval. On the first simulation run we made with **simmm1**, the length of each of the 100 subruns was 2500 and the confidence interval was of length 0.34871. On the last simulation run we made with **simmm1**, the length of each subrun was only 250 (one-tenth that of the first simulation run) and the confidence interval was 1.11709. The error (difference between the true value and the value estimated by the simulation experiment) was only 1.53 percent on the first run but rose to 10.94 percent on the last run. In both cases the true average response time was inside the confidence interval.

Another method sometimes used in place of the method of batched means is called the *method of independent replications*. For this method a number of independent runs are made by using different random number streams on different runs. The runs are made independent by making them very long. Each run is divided into a transient phase and a steady-state phase. For each run the steady-state phase is used to make estimates of the characteristics of interest such as mean response times. These estimates are combined to make the final estimates and a confidence interval for each is calculated using arguments based on the t-distribution. The method of independent replications is described by Welch [Welch 1983].

In the program **simmm1** we use some special properties of the exponential distribution. For an explanation of why the program works see [Morgan 1984]. Let us consider an example. Suppose we choose an average arrival rate λ of 0.8 customers per second and an average service time S of 1 second. This means that the average server utilization is 0.8 by the utilization law $U = \lambda \times S$. MacDougall's algorithm in [MacDougall 1987] recommends a warmup length of 250 (n = 250) and a batch length of 2500 (m = 2500) to start. This warmup length seems to be too short. (MacDougall's algorithm will correct for this.) For the M/M/1 system with $\lambda = 0.8$ and $S = 1$ the true average value of response time is 5 seconds. We display some output from **simmm1** and the exact solution using **mm1**:

```
In[4]:= simmm1[0.8, 1.0, 13, 250, 2500]//Timing
The mean value of time in system at end of warmup is
4.0033
Mean time in system is 4.92449
95 percent confidence interval is
4.75014 to 5.09885

Out[4]= {872.92 Second, Null}

In[5]:= mm1[0.8, 1.0]//Timing
The server utilization is  0.8
The value of Wq is  4.
The value of W is  5.
The average number in the queue is  3.2
The average number in the system is  4.
The average number in a nonempty queue is  5.
The 90th percentile value of q is  10.39720771
The 90th percentile value of w is  11.51292546
```

```
Out[5]= {0.03 Second, Null}


In[4]:= simmm1[0.8, 1.0, 17, 2500, 500]//Timing
The mean value of time in system at end of warmup is
7.6083
Mean time in system is 4.72397
95 percent confidence interval is
4.40889 to 5.03904

Out[4]= {183.17 Second, Null}

In[5]:= simmm1[0.8, 1.0, 31, 10000, 250]//Timing
The mean value of time in system at end of warmup is
5.4389
Mean time in system is 5.54681
95 percent confidence interval is
4.98826 to 6.10535

Out[5]= {123.49 Second, Null}
```

The purpose of printing out the value of mean response time at the end of the warmup period is to determine whether or not it seems likely that the steady-state has been reached. Since the correct value of mean response time is 5.0, the run length of 250 didn't seem to be long enough. But neither did a run of length 2500 where the error rose from 0.9967 (for the run of length 250) to 2.6083 (for the run of length 2500)! A warmup period of 10000 appeared to be adequate. However, the batch runs should have been longer than 250 in our last run as the large confidence interval shows. MacDougall, in Table 4.2 of [MacDougall 1987], claims that to obtain 5% accuracy in the average queueing time (response time minus service time) requires a sample size (run length) of 189774. We had a sample size of 250000 after the warmup in our first run and the estimated average queueing time of 3.92449 is in error by only 1.92%. The error in the average response time is 1.53%. We show the exact values of all the performance measures in the output of the program **mm1**. Note that **mm1** required only 0.03 seconds for the calculation while the first simulation run was 872.92 seconds (14 minutes and 32.92 seconds) long.

Our **simmm1** example illustrates some of the problems of simulation. We will discuss other problems after the following exercise.

Exercise 6.1

Make two M/M/1 simulation runs with **simmm1**, first with a lambda value of 0.9, an average service time of 1.0 seconds, a seed of 11, a warmup value (n) of 1500, and a batch length value (m) of 500. Then repeat the run with all values the same except the batch length (m); make it 2000. Compare the 95 percent confidence intervals for the two runs. (Warning: The first run on my 33 MHz 486 PC took 253.21 seconds and the second 982.4 seconds. If you have a slower computer, such as a 16 MHz 386SX, the two runs could be very long. In this case you may want to take a coffee break or a walk around the block while the computations are made.)

The basic problem in discrete event simulation is that the outputs of a simulator are sequences of random variables rather than the exact performance numbers we would like. The conclusions of a simulation study are based on estimates made from these random variables. Therefore, the estimates themselves are also random variables rather than the performance numbers we want. We usually are interested in estimates of the average values of performance parameters of the computer system under study. For example, we are interested in the average response time of customers in a workload class. If we push n customers of workload class c through the simulator, we obtain the numbers $R_1, R_2, \ldots, R_n$. From these numbers, which are the measured values of the response times for the n customers, the simulator must estimate the average response time for the class. If n is 10000, we may have the simulator ignore the first 1000 of these 10000 numbers to avoid the transient phase and estimate the true value of the average response time R by $\hat{R}$ where

$$\hat{R} = \frac{1}{9000} \times \sum_{i=1001}^{10000} R_i.$$

This is the usual method of estimating an average value; $\hat{R}$ is called the *sample mean* of the numbers $R_{1001}, R_{1002}, \ldots, R_{10000}$. It is important in a simulation study not only to be able to obtain estimates of important parameters from the study, but also to have some sort of assurance that the estimate is close enough to the true value to satisfy the needs of the modeling study. In the program **simmm1** we used the method of batch means to calculate a 95 percent confidence interval for the mean response time. There are a couple of other methods that are sometimes used for this purpose and also help with the problem of determining that the simulation process has reached the steady-state. Unfortunately, both of these methods are rather advanced and thus not easy for beginners to implement. Some simulation languages, such as RESQ, have built-in facilities for both these methods.

The first advanced method is called the *regeneration method.* This method simultaneously solves three problems: (1) the problem of independent runs. (2) the problem of the transient state, and (3) the problem of generating a confidence interval for an estimate. In our discussion of the method of batch means, we neglected to mention the problem of making the batch runs independent. What tends to keep them from being independent is the correlation between successive customers. If one customer has a very long response time because of long queues at the service centers, then immediately succeeding customers tend to have long response times as well; of course, if a customer has a short response time, then immediately succeeding customers tend to have short response times, too. The batch runs are approximately independent if each of them is sufficiently long, however. The regeneration method automatically generates independent subruns. The regenerative method also solves the problem of the transient state. Finally, the regenerative method supplies a technique for generating confidence intervals. With these three advantages one might suppose that everyone should use the regenerative method. Unfortunately, there are disadvantages for the regenerative method, too. The method does not apply to all simulation models, although it does apply to the simulation of most computer systems. In addition it is much more complex to set up properly and more difficult to program.

The regeneration method depends upon the existence of *regeneration* or *renewal* points. At each such point future behavior of the simulation is independent of past behavior and, in a probabilistic sense, restarts or regenerates its behavior from that point. Eventually the system returns to the same regeneration point or state in what is called a *regeneration cycle*. The regeneration cycles are used as subruns for the simulation study. Since each regeneration point represents identical simulation model states, the behavior of the system during one cycle is independent of the behavior in another cycle, so the subruns are independent. The bias due to the initial conditions also disappears. An example of a regeneration point for the M/M/1 queueing model is the initial state in which the system is empty and the first customer to enter the system will appear in Δt seconds where Δt is a random number from an exponentially distributed stream with average value $1/\lambda$. The first regeneration cycle ends the next time the simulated system again reaches the empty state.

In Section 3.3.2 of [Bratley, Fox , and Schrage 1987], the authors discuss regenerative methods, provide an algorithm for using the regeneration method, and give a list of pros and cons of the regenerative method. One of the cons is that the regeneration cycles may be embarrassingly long. Although Bratley *et al.* didn't mention it, there may be extremely short regeneration cycles as well. Another problem is in setting up regeneration points to begin a simulation. This can be a real challenge. The regeneration method is not recommended for beginners.

There is also a discussion of the spectral method in [Bratley, Fox, and Schrage 1987]. The spectral method is supported by the RESQ programming language and examples of its use are given in [MacNair and Sauer 1985]. The method does provide confidence intervals for steady-state averages. In addition, MacNair and Sauer claim:

> A sequential stopping rule is also available with the spectral method. A significant advantage of the spectral method over independent replications is that we can make a single (long) simulation run instead of multiple (shorter) runs. Therefore we do not need to be as concerned about the effects of the choice of the initial state. The spectral method applies to equilibrium behavior of all models simulated using extended queueing networks, not just those with regenerative properties.

Bratley *et al.* [Bratley, Fox, and Schrage 1987] discuss other advanced methods, which they call autoregressive methods. These methods are not widely used and Bratley *et al.* do not present an optimistic portrayal of their use. In fact, they end Section 3.3 with the statement:

> To construct a confidence interval, one can pretend (perhaps cavalierly) that this approximate t-distribution is exact. Law and Kelton (1979) replace any negative $\bar{R}_s$ by 0, though for typical cases this seems to us to have a good rationale only for small and moderate s. With this change, they find that the confidence intervals obtained are just as accurate as those given by the simple batch means method. Duket and Pritsker (1978), on the other hand, find spectral methods unsatisfactory. Wahba (1980) and Heidelberger and Welch (1981a) aptly criticize spectral-window approaches. They present alternatives based on fits to the logarithm of the periodogram. Heidelberger and Welch (1981a, b) propose a regression fit, invoking a large number of asymptotic approximations. They calculate their periodogram using batch means as input and recommend a heuristic sequential procedure that stops when a confidence interval is acceptably short. Heidelberger and Welch (1982) combine their approach with Schruben's model for initializa-

> tion bias to get a heuristic, composite, sequential procedure for running simulation. Because the indicated coverage probabilities are only approximate, they checked their procedure empirically on a number of examples and got good results. Despite this, we believe that spectral methods need further study before they can be widely used with confidence. For sophisticated users, they may eventually dominate batch means methods but it seems premature to make a definite comparison now.

One of the major challenges in writing a simulator is in generating the required streams of random numbers. Even if you use a simulation modeling package that provides facilities for random number generation you should test the output of such streams to be sure they are correct.

Anyone who considers arithmetical methods of producing random digits is, of course, in a state of sin.
John von Neumann

Every random number generator will fail in at least one application.
Donald E. Knuth

6.3.1 Random Number Generators

We saved until last the problem of generating random numbers. We have already described how to generate random numbers with an exponential distribution using *Mathematica*. The algorithm we used depended upon the fact that *Mathematica* has a random number generator **Random**, which can be used to generate a sequence of random numbers that are uniformly distributed on the interval 0 to 1. Such a random number generator, called a *uniform random number generator*, is the key to generating a random sequence with any given kind of distribution. Algorithms exist for converting a sequence of uniform random numbers to a sequence of random numbers with any given probability distribution. A good uniform random number generator should be able to produce a very long sequence of statistically independent random numbers, uniformly distributed on the interval from 0 to 1. As Park and Miller point out in their paper [Park and Miller 1988], many uniform random number generators in subroutine libraries of computer installations as well as in computer science textbooks are flawed. The authors say:

> Many generators have been written, most of them have demonstrably non-random characteristics, and some are embarrassingly bad.

The random generator most universally condemned by experts is **RANDU**, the generator that appears in the IBM System/360 Scientific Subroutine Package (a package well thought of except for this program). Knuth in his outstanding book [Knuth 1981] says

> Unfortunately, quite a bit of published material in existence at the time this chapter was written recommends the use of generators that violate the suggestions above; and the most common generator in actual use, **RANDU,** is really horrible (cf. Section 3.3.4).

Knuth mentions this program in a pejorative manner in several other places in his book.

The most common random number generators are *linear congruential generators* that work as follows: Given a positive integer m and an initial *seed* z_0, with $0 \le z_0 < m$, the sequence $z_0, z_1, z_2, \ldots$ is generated with $z_{n+1} = az_n + b \bmod m$ where a and b are integers less than m. The integer a is called the *multiplier*, and is in the range $2, 3, \ldots, m-1$, b is called the *increment*, and m the *modulus*. In the formula for generating the next random number, "mod m" means to take the remainder upon division by m. Thus, if m is 13, then 27 mod m is 1.

Park and Miller recommend a standard uniform random number generator based on a linear congruential generator with increment zero. They also recommend that the modulus m be a large *prime* integer. (Recall that a positive integer m is prime if the only positive integers that divide it evenly are 1 and m. By convention, 1 is not considered a prime number so the sequence of prime numbers is $2, 3, 5, 7, 11, 13, 17, \ldots$.) Their algorithm is begun by choosing a seed z_1 and generating the sequence $z_1, z_2, z_{3, \ldots}$ by the formula $z_{n+1} = a \times z_n \bmod m$ for $n = 1, 2, 3, \ldots$. Finally, each z_n is converted into a number between zero and one by dividing by m which yields a new sequence $u_1, u_2, u_3, \ldots$ where $u_n = z_n/m$. Park and Miller refer to this algorithm as the *Lehmer generator*. The numbers m and a must be chosen very carefully to make the Lehmer generator work properly.

We implement Lehmer's algorithm in the *Mathematica* program **uran**, which uses the program **ran**. In the program **ran** we generate a random sequence of integers but do not divide each by m so **ran** is not a uniform random number generator; it generates a sequence of integers between 1 and m. The *Mathematica* program **uran** is a uniform random number generator. The programs **ran** and **uran** are part of the package **work.m** and are listed below:

```
ran[a_Integer, m_Integer, n_Integer, seed_Integer]:=
Block[{i},
output =Table[0, {n}];
output[[1]]=Mod[seed, m];
For[i = 2, i<=n, i++,
output[[i]] = Mod[a output[[i-1]], m]];
Return[output];
]
uran[a_Integer, m_Integer, n_Integer, seed_Integer]:=
Block[{i},
random = ran[a, m, n, seed];
output =Table[0, {n}];
output[[1]]=Mod[seed, m]/m;
For[i = 2, i<=n, i++,
output[[i]] = random[[i]]/m];
Return[output];
```

All linear congruential generators are periodic; that is, after a certain number of iterations the generator repeats itself. Let us illustrate by an example from [Park and Miller 1988]. Suppose we choose the random Lehmer generator with the multiplier $a = 6$ and modulus $m = 13$. Then, if the initial seed is 2, the Lehmer generator yields the sequence (before dividing by 13) of $2, 12, 7, 3, 5, 4, 11, 1, 6, 10, 8, 9, 2, \ldots$ After the second 2 the sequence repeats itself. The choice of any other initial seed would yield a circular shift of the above sequence. This generator is a *full period generator*, that is, it yields all the numbers from 1 through 12 exactly once in each period. The multiplier $a = 5$ in the above example yields a Lehmer generator with period of only four; it is not a full period generator. We demonstrate these properties with the Mathematica program **ran**:

```
In[6]:= ran[6, 13, 20, 2]

Out[6]= {2, 12, 7, 3, 5, 4, 11, 1, 6, 10, 8, 9, 2, 12,
```

```
7, 3, 5, 4, 11, 1}

In[7]:= ran[5, 13, 20, 2]

Out[7]= {2, 10, 11, 3, 2, 10, 11, 3, 2, 10, 11, 3, 2,
10, 11, 3, 2, 10, 11, 3}

In[5]:= N[uran[6, 13, 5, 2], 8]

Out[5]= {0.15384615, 0.92307692, 0.53846154,
0.23076923, 0.38461538}
```

The statement on line 5 shows how the *Mathematica* program **uran** can be used to generate uniform random variables on the interval between zero and one.

Exercise 6.2

Consider the Lehmer generator with $m = 13$. We saw that with the multiplier $a = 6$ we have a full period generator, while the multiplier $a = 5$ yields a generator with a period of only 4. Test all the other multipliers between 2 and 12 to see which give you a full period Lehmer generator.

Knuth [Knuth 1981] discusses how to choose the parameters of a linear congruential generator to obtain a full period. He considers generators with $b = 0$ as a special case. The solution for this case is given by Theorem B on page 19 of the Knuth book. A linear congruential generator with $b = 0$ is called a *multiplicative linear congruential generator.* Every full period linear congruential generator produces a fixed circular list; the initial seed determines the starting point on this list for the output of any particular run.

Another desirable property of a random number generator is that the output be random. As Gardner shows in [Gardner 1989, Gardner 1992], the exact meaning of random is difficult to define. Loosely speaking, the output of a random number generator is random if it appears to be so. Statistical tests have been designed to test this property because humans cannot make good judgments about randomness. Knuth [Knuth 1981] has a long, difficult section with the title "What is a random sequence?" It turns out that, if a sequence is random, then subsequences must exist that appear to be very nonrandom. That is, sequences such as $1, 2, 3, 4, 5, 6, 7, 8, 9, 0$. In practice we must depend upon statistical tests to decide whether or not a random number generator yields random output. Some choices of a and m for the Lehmer generator yield sequences that are more random than others. It is not easy to choose the combinations of a and m for a Lehmer generator that will generate satisfactory random output. For their minimal standard random number generator Park and Miller recommend the multiplier $a = 16807$ with the modulus $m = 2147483647$. They chose $m = 2^{31} - 1 = 2147483647$ because it is a large prime. For this value of m there are more than 534 million values of a that make the generator a full period generator. Extensive testing has been performed which suggests that the combination of a and m recommended by Park and Miller does yield a truly random full period sequence. Being "truly random" means that it has passed the statistical tests that are used to determine randomness or lack of it. The Park and Miller minimal standard random number generator has been implemented successfully on a number of computer platforms.

From a uniform random number generator, which generates a sequence $u_1, u_2, u_3, \ldots$ where each u_n is between zero and one, it is possible to generate a sequence with any probability distribution desired. Knuth [Knuth 1981] includes algorithms for most distributions of interest to those modeling computer systems. Some of the algorithms are somewhat complex but the algorithm for generating an exponentially distributed random sequence is very straightforward. One can generate an exponentially distributed random sequence with average value x by calculating $b_n = -x \times \log u_n$ for each n where the log function is the natural logarithm, that is, the logarithm to the base e where e is approximately 2.718281828. The *Mathematica* program **rexpon** can be used to generate an exponential random sequence.

```
rexpon[a_Integer, m_Integer, n_Integer, seed_Integer,
mean_Real]:=
Block[{i,random, output},
random = uran[a, m, n, seed];
output =Table[0, {n}];
For[i =1, i<=n, i++,
output[[i]] = - mean Log[random[[i]]]];
Return[N[output, 6]];

In[14]:= rexpon[6, 13, 10, 2, 3.5]

Out[14]= {6.55131, 0.280149, 2.16664, 5.13218,
3.34429, 4.12529, 0.584689,

>     8.97732, 2.70616, 0.918275}
```

```
In[15]:= Apply[Plus, %]/10

Out[15]= 3.47863
```

In the preceding example we generated an exponential random sequence of length 10 with mean (average) 3.5. Note that the average of these numbers is not exactly 3.5 but is fairly close to it. Of course if the desired average value of the exponentially distributed random numbers is x, we can generate one such number by the statement $s = -x$ Log[Random[]].

Mathematica has the capability of generating random sequences directly with any random variable that is supported by *Mathematica* such as the continuous distributions in the package **Statistics‘ContinuousDistributions’**. We demonstrate how to generate a sequence of exponential random variates with a mean of 3.5 in the following *Mathematica* run:

```
In[3]:= <<Statistics'ContinuousDistributions'

In[4]:= table1 = Table[Random[
ExponentialDistribution[1/3.5]], {20}];

In[5]:= Mean[table1]

Out[5]= 3.56487

In[6]:= table1 = Table[Random[
        ExponentialDistribution[1/3.5]], {20}];

In[7]:= Mean[table1]

Out[7]= 4.62718

In[8]:= table1 = Table[Random[
                ExponentialDistribution[1/3.5]], {20}];

In[9]:= Mean[table1]

Out[9]= 2.86325

In[10]:= table1 = Table[Random[
ExponentialDistribution[1/3.5]], {10000}];
```

```
In[11]:= Mean[table1]

Out[11]= 3.53028

In[12]:= Variance[table1]
Out[12]= 12.73

In[13]:= 3.5^2

Out[13]= 12.25
```

Note that for small samples, such as 20, the mean was not always close to 3.5, but for a sample of size 10000, both the mean and variance were fairly close to the underlying distribution. (The variance for an exponential random variable is the square of its mean, so, if the mean is 3.5, the variance should be 12.25.)

Marsaglia is one of the leaders in random number generation. In his keynote address "A Current View of Random Number Generators" for the Computer Science and Statistics: 16th Symposium on the Interface, Atlanta, 1984, which is published as [Marsaglia 1985] he made some important remarks. He said, in the abstract:

> The ability to generate satisfactory sequences of random numbers is one of the key links between Computer Science and Statistics. Standard methods may no longer be suitable for increasingly sophisticated uses, such as in precision Monte Carlo studies, testing for primes, combinatorics or public encryption schemes. This article describes stringent new tests for which standard random number generators: congruential, shift-register and lagged-Fibonacci, give poor results, and describes new methods that pass the stringent tests and seem more suitable for precision Monte Carlo use.

He begins his address on a conciliatory note:

> 1. INTRODUCTION
> Most computer systems have random number generators available, and for most purposes they work remarkably well. Indeed, a random number generator is much like sex: when it's good it's wonderful, and when it's bad it's still pretty good.

In Part 2 Marsaglia becomes a little less sanguine:

> 2. SIMPLE GENERATORS: CONGRUENTIAL
> These generators use a linear transformation on the ring of reduced residues of some modulus m, to produce a sequence of integers: $x_1, x_2, x_3, \ldots$ with $x_n = ax_{n-1} + b \bmod \mathrm{m}$. They are mostly widely used RNG's, and they work remarkably well for most purposes. But for some purposes they are not satisfactory; points in n-space produced by congruential RNG's fall on a lattice with a huge unit cell volume, m^{n-1}, compared to the unit cell volume of 1 that would be expected from random points with coordinates constrained to be integers. Details are in [9,10]. Congruential RNG's perform well on many of the stringent tests described below, but not on all of them.

Marsaglia then describes some of the other common random number generators, some new generators, some new, more stringent tests, and the results of applying the tests to old and new random number generators. He concludes with the following paragraph:

> Based on the above discussion, my current view of RNG's may be summarized with the following bottom line: Combination generators seem best; congruential generators are liked, but not well-liked; shift-register and lagged-Fibonacci generators using no-carry add are no good; avoid no carry add; lagged Fibonacci generators using + or - pass most of the stringent tests, and all of them if the lag is long enough, say 607 or 1279; Lagged-Fibonacci generators using multiplication on odd integers mod 2^{32} pass all the tests; combination generators seem best—if the numbers are not random, they are at least higgledy piggledy.

In 1991, Marsaglia and Zaman in [Marsaglia and Zaman 1991] announced a breakthrough in random number generators. Their new generators are called *add-with-carry* and *subtract-with-borrow*. In [Marsaglia and Zaman 1992], Marsaglia and Zaman announced the availability of ULTRA, a random number generator based on their subtract-with-borrow algorithm. They provide an assembler program for 80x86 processors as well as a version written in C. The code is free to anyone who sends them a DOS floppy. Marsaglia and Zaman claim that:

"ULTRA has a period of some 10^{366} and that every possible m-tuple, from pairs, 3-tuples, 4-tuples up to 37-tuples, can appear. Statistical tests show that those m-tuples appear with frequencies consistent with underlying probability theory."

If you read [Knuth 1981] you will be amazed by the number of tests for randomness he provides. However, if you do a simulation study you may be tempted to skip the testing of your random number generator. This would be a mistake. Jon Bentley, the author of the regular column *Software Exploratorium* in *UNIX Review,* in [Bentley 1992] discusses the use of a random number generator to study the approximate solution to the traveling-salesman problem. He uses a random number generator recommended by Knuth in Algorithm A and implemented by Program A written in Knuth's MIXAL on page 27 of [Knuth 1981]. Bentley tested his version of the program more thoroughly than Knuth did and discovered that, for his application, Knuth's recommendation wouldn't work! If he had not done the extensive testing he may not have discovered the error for some time. Bentley found a modification to the algorithm based on some of Knuth's recommendations that does work satisfactorily. In his column Bentley gave the following exercise:

> **Exercise 12**. Implement Knuth's generator verbatim from the Further Reading. Does it display similar problems when used with **fortune**? If so, trace the problems.

In Exercise 12, **fortune** refers to a program that reads a file of one-line quotations and prints one at random. The generator referred to is a FORTRAN program on page 171 of [Knuth 1981]. The answer to Exercise 12 provided by Bentley is:

> **12**. Knuth's implementation was also flawed: it never chose the sixth line in the file. I found that for every seed less than 100,000, whenever the sixth integer generated is congruent to 0 modulo 6, the ninth integer is congruent to 0 modulo 9 (and thus the ninth line is chosen rather than the sixth).

Knuth is one of the most admired computer scientists of our time. His book [Knuth 1981] is the standard reference on random number generation. His final advice in the SUMMARY for the chapter RANDOM NUMBERS includes the following statements:

> The authors of many contributions to the science of random number generation were unaware that particular methods they

> were advocating would prove to be inadequate. Perhaps further research will show that even the random number generators recommended here are unsatisfactory; we hope this is not the case, but the history of the subject warns us to be cautious. The most prudent policy for a person to follow is to run each Monte Carlo program at least twice using quite different sources of random numbers, before taking the answers of the program seriously; this not only will give an indication of the stability of the results, it also will guard against the danger of trusting in a generator with hidden deficiencies. (Every random number generator will fail in at least one application.)

Peterson reports in [Peterson 1992] that Alan M. Ferrenberg, a computational physicist at the University of Georgia, discovered that the random number generator developed by Marsaglia and Zaman can yield incorrect results under certain circumstances. Ferrenberg simulated a two-dimensional Ising model for which he knew the correct answer using the Marsaglia and Zaman algorithm for generating the random numbers and got an incorrect result. When he used a linear congruential generator for the simulation he got much more accurate results. Ferrenberg's experience is in agreement with Knuth's statement, "Every random number generator will fail in at least one application."

We use the program **chisquare** to test a random sequence to see if it has an exponential distribution with a given mean using the chi-square test. If you have taken a statistics course of any kind you are probably familiar with the chi-square test. (Warning: The program **chisquare** only tests the sequence to see if it has an exponential distribution. That is, **chisquare** will tell you whether or not a given sequence appears to have an exponential distribution with a given mean. It will not test for any other distribution such as normal or uniform.)

```
   chisquare[alpha_, x_, mean_]:=
   Block[{n, y, xbar, x25, x50, x75, o, e, m, first},
   chisdist = ChiSquareDistribution[3];
   n= Length[x];
   y = Sort[x];
   (* We calculate the quartile values assuming x is
exponential. *)
   x25 = - mean Log[0.75];
   x50 = -mean Log[0.5];
   x75 = -mean Log[0.25];
   o = Table[0, {4}];
```

```
   o[[1]] = Length[Select[y, # <= x25 &]];
   o[[2]] = Length[Select[y, x25 < # && # <= x50 &]];
   o[[3]] = Length[Select[y, x50 < # && # <= x75 &]];
   o[[4]] = Length[Select[y, # > x75 &]];
   (* o is the observed number in each quarter defined
by *)
   (* the quartiles. *)
   m = n/4;
   e = Table[m, {4}];
   (* e is the expected number in each quarter. One-
fourth *)
   (* in each. *)
   first = ((o - e)^2)/m;
   chisq = N[Apply[Plus,first], 6];
   (* This is the chisq value. *)
   q = CDF[chisdist, chisq];
   (* q is the probability that any observed chisq
value *)
   (* will not exceed the value just observed *)
   (* if x is exponential. *)
   p = 1 - q;
   (* p is the probability any value of chisq will be
*)
   (* greater than or equal to that just observed *)
   (* if x is exponential. *)
   Print["p is ", N[p, 6]];
   Print["q is ", N[q, 6]];
   If[p < alpha/2, Return[Print["The sequence fails
because chisq is too large."]]];
   If[q < alpha/2, Return[Print["The sequence fails
because chisq is too small."]]];
   If[p >= alpha/2 && q >= alpha/2, Return[Print["The
sequence passes the test."]]
   ]
```

The program **chisquare** applies the chi-square test to the random sequence x. The chi-square test is a goodness-of-fit test. Such a statistical test is a special case of a hypothesis test. A hypothesis test works by attempting to show that a null hypothesis is not reasonable at the α level of significance where α is usually taken to be 5% (0.05) or 1% (0.01). The null hypothesis in **chisquare** is that the random sequence x has an exponential distribution with a given average value *mean*.

To apply the chi-square test to a sequence of random numbers $x_1, x_2, x_3, \ldots, x_n$ we must assume that n is large and the numbers are independent (at least they must appear to be). (There are other tests that can be used to measure the independence.) We assume that each number fits into one of k categories. We use the symbol O_i for the number of the random numbers that fall into category i, for $i = 1, 2, \ldots, k$. Then we calculate the expected number of the random numbers that would fall into each category given that the sequence has the assumed probability distribution. We use the symbol E_i for the expected number for $i = 1, 2, \ldots, k$. We then calculate the chi-square value, *chisq*, as a measure of the deviation of the observed sequence from the assumed exact distribution where

$$\text{chsq} = \frac{(O_1 - E_1)^2}{E_1} + \frac{(O_2 - E_2)^2}{E_2} + \ldots + \frac{(O_k - E_k)^2}{E_k}.$$

Each numerator in the sum for chisq measures the square of the difference between the observed and expected number in a category; the number in each denominator scales the squared value. Fortunately, for large n, the distribution of chisq approaches the well-known probability distribution called the *chi-square* distribution. The chi-square distribution is completely characterized by one integer parameter called the *degrees of freedom.*

In the program **chisquare** $k = 4$. We calculate the three numbers x25, x50, and x75 which define four intervals of the real line in such a way that, if the random sequence has an exponential distribution with mean value *mean*, then one-fourth of the sequence will fall into each interval. Since we assume we know the mean of the sequence, by the rules for calculating number of degrees of freedom of the chi-square distribution approximating chisq, it has $k - 1 = 3$ degrees of freedom. If our null hypothesis had been merely that the sequence was exponential so that we must estimate the mean from the data we would lose another degree of freedom so that chisq would be approximated by a chi-square distribution with 2 degrees of freedom. We now provide some output from **chisquare** that shows some tests of exponential random numbers generated by **rexpon** and by *Mathematica* using **Random**. The *Mathematica* package **work.m** was loaded before the statements below were executed using Version 2.0 of *Mathematica.* SeedRandom yields different values for other versions of *Mathematica* so you may get somewhat different results from if you use a version of Mathematica other than 2.0.

```
In[5]:= y = rexpon[16807, 2147483647, 5000, 2, 3.5];//
Timing

Out[5]= {166.21 Second, Null}

In[6]:= Mean[y]

Out[6]= 3.54594

In[7]:= chisquare[0.02, y, 3.5]
p is 0.989953
q is 0.010047
The sequence passes the test.

In[8]:= SeedRandom[2]

In[9]:= x = Table[Random[ExponentialDistribution[1/
3.5]], {5000}];//Timing

Out[9]= {11.55 Second, Null}

In[10]:= chisquare[0.02, x, 3.5]
p is 0.0111519
q is 0.988848
The sequence passes the test.

In[11]:= Mean[x]

Out[11]= 3.54394

In[12]:= SeedRandom[23]

In[13]:= x = Table[Random[ExponentialDistribution[1/
3.5]], {5000}];//Timing

Out[13]= {12.16 Second, Null}

In[14]:= Mean[x]

Out[14]= 3.52034
In[15]:= chisquare[0.02, x, 3.5]
p is 0.946125
```

```
q is 0.0538745
The sequence passes the test.

In[17]:= y = rexpon[16807, 2147483647, 5000, 23,
3.5];//Timing

Out[17]= {177.9 Second, Null}

In[18]:= chisquare[0.02, y, 3.5]
p is 0.473991
q is 0.526009
The sequence passes the test.

In[19]:= y = rexpon[16807, 2147483647, 5000, 37,
3.5];//Timing

Out[19]= {177.4 Second, Null}
In[20]:= chisquare[0.02, y, 3.5]
p is 0.0860433
q is 0.913957
The sequence passes the test.

In[5]:= SeedRandom[47]

In[6]:= y = Table[Random[ExponentialDistribution[1/
10]], {5000}];

In[8]:= chisquare[0.02, y, 20]
p is 0.
q is 1.
"The sequence fails because chisq is too large."
```

Although **rexpon** uses the Park and Miller minimal standard random number generator, which they claim is very efficient, it required 166.21 seconds to generate 5000 exponential random variates compared to 11.55 seconds required to produce them using the *Mathematica* **Random** function. The program **chisquare** rejects the sequence if p is less than half of alpha or q is less than half of alpha. We calculate p as the probability that, if the null hypothesis is true, a value of *chisq* as large or larger than the one observed would be observed. Similarly, q represents the probability that an observed value of *chisq* smaller than that observed would occur. We have followed Knuth's recommendation of testing each random number generator at least three times with different seeds. Both random number generators pass all the tests with an alpha of 0.02 (two percent). Some authorities would not reject the sequence based on q being less than one half of alpha but would reject only if p is less than alpha. We follow Knuth's recommendation in choosing success or failure of the sequence in **chisquare**.

Exercise 6.3

Load the *Mathematica* package **work.m** and use **chisquare** to test the sequence generated by the *following Mathematica* statements to see if it is a random sample from an exponential distribution with mean 10. Use 0.02 as the alpha value.

```
In[4]:= SeedRandom[47]

In[5]:= y = Table[Random[ExponentialDistribution[1/
10]], {1000}];
```

6.4 Simulation Languages

Except for very trivial models, simulation involves computer computation and therefore some programming language must be used to code the simulator. There are three kinds of languages that can be used for computer performance analysis simulation models:

1. General programming languages such as Pascal, FORTRAN, or C++.
2. General purpose simulation languages such as GPSS or SIMSCRIPT II.5.
3. Special purpose simulation languages such as PAWS, SCERT II, and RESQ.

Simulation languages of the third type are specifically designed for analyzing computer systems. These languages have special facilities that make it easier to construct a simulator for a modeling study of a computer system. The modeler is thus relieved of a great deal of complex coding and analysis. For example, such languages can easily generate random number distributions of the kind usually used in models of computer systems. In addition Type 3 languages make it easy to simulate computer hardware devices such as disk drives, CPUs, and channels as part of a computer system simulation. Some languages, such as RESQ, also allows advanced methods for controlling the length of a simulator run such as the regeneration method, running until the confidence interval for an estimated performance parameter is less than a critical value, etc. Type 3 languages are more expensive, in general, than Type 1 or Type 2 languages, as one would expect, but provide a savings in the time to construct a simulator. Of course there is a learning curve for any new language; it might be necessary to attend a class to attain the best results.

Type 2 programming languages provide a number of features needed for general purpose simulation but no special features for modeling computer systems as such. Therefore, it is easier to develop a simulator with a Type 2 programming language than with a Type 1 general purpose language, but not as easily as with a Type 3 language.

Type 1 languages should be used for constructing a simulator of a computer system only if (a) the simulator is to be used so extensively that efficiency is of paramount importance, (b) personnel with the requisite skills in statistics and coding are available to construct the model, and (c) a simple technique for prototyping the simpler versions of the simulator is available to assist in validating the simulator.

Bratley *et al.* [Bratley, Fox, and Schrage, 1987] provide examples of simulators written in Type 1 languages (Fortran, and Pascal) as well as Type 2 languages (Simscript, GPSS, and Simula). They also warn:

> Finally, the best advice to those about to embark on a very large simulation is often the same as Punch's famous advice to those about to marry: Don't!

MacNair and Sauer [MacNair and Sauer 1985] provide a number of computer modeling examples using simulation written in the Type 3 language RESQ.

6.5 Simulation Summary

Simulation is a powerful modeling techniques but requires a great deal of effort to perform successfully. It is much more difficult to conduct a successful modeling study using simulation than is generally believed.

Challenges of modeling a computer system using simulation include:

1. Determining the goal of the study.
2. Determining whether or not simulation is appropriate for making the study. If so, determine the level of detail required. It is important to schedule sufficient time for the study.

3. Collecting the information needed for conducting the simulation study. Information is needed for validation as well as construction of the model.
4. Choosing the simulation language. This choice depends upon the skills of the people available to do the coding.
5. Coding the simulation, including generating the random number streams needed, testing the random number streams, and verifying that the coding is correct. People with special skills are needed for this step.
6. Overcoming the special simulation problems of determining when the simulation process has reached the steady-state and a method of judging the accuracy of the results.
7. Validating the simulation model.
8. Evaluating the results of the simulation model.

A failure of any one of these steps can cause a failure of the whole effort.

Simulation is the only tool available for modeling computer hardware that does not yet exist and thus is of great importance to computer designers. It also plays a leading role in analyzing the performance of complex communication networks. Fortier and Desrochers [Fortier and Desrochers 1990] describe how the MATLAN simulation modeling package can be used to analyze local area networks (LANs).

> **bench mark.** *A surveyor's mark made on some stationary object of previously determined position and elevation, and used as a reference point in tidal observations and surveys.*
>
> *American Heritage Dictionary 1981*

6.6 Benchmarking

We discussed benchmarking briefly in Chapters 1 and 2. There are actually two basically different kinds of benchmarking. The first kind is defined by Dongarra *et al.* [Dongarra, Martin, and Worlton 1987] as "Running a set of well-known programs on a machine to compare its performance with that of others." Every computer manufacturer runs these kinds of benchmarks and reports the results for each announced computer system. The second kind is defined by Artis and Domanski [Artis and Domanski 1988] as "a carefully designed and structured experiment that is designed to evaluate the characteristics of a system or subsystem to perform a specific task or tasks." The first kind of benchmark is represented by the Whetstone, Dhrystone, and Linpack benchmarks. According to [Artis and Domanski 1988] the second kind of benchmark can be used as follows:

1. A benchmark may be used to evaluate the capability of alternative systems to process a specific load to evaluate the cost performance levels of competing hardware proposals.
2. A benchmark may be used to certify the functionality and performance of critical applications after significant modifications have been made to hardware and/or software configurations.
3. A benchmark may be used to stress-test hardware or software during acceptance periods.
4. A benchmark may be used to provide "yardstick" measures of resource consumption to calibrate accounting rates for new processors or configurations.
5. A benchmark may be used to certify the performance of prototype applications.
6. A benchmark may be used to fulfill legislated or policy requirements for "fairly" selecting new hardware or software systems.
7. A benchmark may be used to provide a learning experience.

The Artis and Domanski kind of benchmark is the type one would use to model the workload on your current system and run on the proposed system. It is the most difficult kind of modeling in current use for computer systems. Before we discuss the Artis and Domanski type of benchmark, we will discuss the first type of benchmark, the kind that is called a *standard* benchmark. We have previously mentioned some of the standard benchmarks, including the Dhrystone benchmark, in Chapter 2.

In the very early days of computers, the speed of different machines was compared using main memory access time, clock speed, and the number of CPU clock cycles needed to perform the addition and multiply instructions. Since most programming in those days was done either in machine language or assembly language, in principle, programmers could use this information plus the cycle times of other common instructions to estimate the performance of a new machine.

The next improvement in estimating computer performance was the Gibson Mix provided by J. C. Gibson of IBM and formally described in [Gibson 1970]. Gibson ran some dynamic instruction traces on a selection of programs written for the IBM 650 and 704 computers. From these traces he was able to calculate what percent of instructions were of various types. For example, he found that Load/Store instructions accounted for 31.2% of all instructions executed and Add/Subtract accounted for 6.1%. From the percentage of each instruction used and the execution time of each instruction, it is possible to compute the average execution time of an instruction and thus the average execution rate. In his excellent historical paper [Serlin 1986] Serlin shows how the Gibson Mix could be used to estimate the MIPS for a 1970-vintage Supermini computer. Serlin also points out that the Gibson Mix was part of industry lore in 1964, although Gibson did not formally publish his results until 1970 and this only in an IBM internal report.

It was quickly discovered that the Gibson Mix was not representative of the work done on many computer systems and did not measure the ability of compilers to produce good optimized code. These concerns led to the development of some standard synthetic benchmarks.

As Engberg says [Engberg 1988] about synthetic benchmarks:

> These are load generators, scripted to mirror the resource consumption patterns of a given workload. These artificial benchmarks can be applied to a specific system in an attempt to measure its impact on that system's performance.

Thus synthetic benchmarks do not do any useful calculations, unlike the Linpack benchmark, which is a collection of Fortran subroutines for solving a system of linear equations. Results of the Linpack benchmark are given in terms of Linpack MFLOPS.

The two best known synthetic benchmarks are the Whetstone and the Dhrystone. The Whetstone benchmark was developed at the National Physical Laboratory in Whetstone, England, by Curnow and Wichman in 1976. It was designed to measure the speed of numerical computation and floating-point operations for midsize and small computers. Now it is most often used to rate the floating-point operation of scientific workstations. My IBM PC compatible 33 MHz 486 has a Whetstone rating of 5,700K Whetstones per second. According to [Serlin 1986] the HP 3000/930 has a rating of 2,841K Whetstones per second, the IBM 4381-11 has a rating of approximately 2,000K Whetstones per second, and the IBM RT PC a rating of 200K Whetstones per second.

The Dhrystone benchmark was developed by Weicker in 1984 to measure the performance of system programming types of operating systems, compilers, editors, etc. The result of running the Dhrystone benchmark is reported in Dhrystones per second. Weicker in his paper [Weicker 1990] describes his original benchmark as well as Versions 1.1 and 2.0. Whetstones per second is often converted into MIPS or millions of instructions per second. The MIPS usually reported are relative VAX MIPS, that is, MIPS calculated relative to the VAX 11/780, which was once thought to be a 1 MIPS machine but is now generally believed to be approximately a 0.5 MIPS machine. By this we mean that for most programs run on the VAX 11/780 it executes approximately 500,000 instructions per second. Weicker [Weicker 1990] not only discusses his Dhrystone benchmark but also discusses the Whetstone, Livermore Fortran Kernels, Stanford Small Programs Benchmark Set, EDN Benchmarks, Sieve of Eratosthenes, and SPEC benchmarks. Weicker also says:

> It should be apparent by now that with the advent of on-chip caches and sophisticated optimizing compilers, small benchmarks gradually lose their predictive value. This is why current efforts like SPEC's activities concentrate on collecting large, real-life programs. Why, then, should this article bother to characterize in detail these "stone age" benchmarks? There are several reasons:
>
> (1) Manufacturers will continue to use them for some time, so the trade press will keep quoting them.
> (2) Manufacturers sometimes base their MIPS rating on them. An example is IBM's (unfortunate) decision to base the published (VAX-relative) MIPS numbers for the IBM 6000 workstation on the old 1.1 version of Dhrystone. Subsequently, DEC and Motorola changed the MIPS computation rules for their competing products, also basing their MIPS numbers on Dhrystone 1.1.
> (3) For investigating new architectural designs—via simulations, for example—the benchmarks can provide a useful first approximation.
> (4) For embedded microprocessors with no standard system software (the SPEC suite requires Unix or an equivalent operating system), nothing else may be available.

Weickert's paper is one of the best summary papers available on standard benchmarks.

According to QAPLUS Version 3.12, my IBM PC 33 MHz 486 compatible executes 22,758 Dhrystones per second. According to [Serlin 1986] the IBM 3090/200 executes 31,250 Dhrystones per second, the HP3000/930 executes 10,000 Dhrystones per second, and the DEC VAX 11/780 executes 1,640 Dhrystones per second, with all figures based on the Version 1.1 benchmark. However, IBM calculates VAX MIPS by dividing the Dhrystones per second from the Dhrystone 1.1 benchmark by 1,757; IBM evidently feels that the VAX 11/780 is a 1,757 Dhrystones per second machine. The Dhrystone statistics on the VAX 11/780 are very sensitive to the software in use. Weicker [Weicker 1990] reports that he obtained very different results running the Dhrystone benchmark on a VAX 11/780 with Berkeley UNIX (4.2) Pascal and with DEC VMS Pascal (V.2.4). On the first run he obtained a rating of 0.69 native MIPS and on the second run a rating of 0.42 native MIPS. He did not reveal the Dhrystone ratings.

Standard benchmarks are useful in providing at least ballpark estimates of the capacity of different computer systems. However, there are a number of problems with the older standard benchmarks such as Whetstone, Dhrystone, Linpack, etc. One problem is that there are a number of different versions of these benchmarks and vendors sometimes fail to mention which version was used. In addition, not all vendors execute them in exactly the same way. That is apparently the reason why Checkit, QAPLUS, and Power Meter report different values for the Whetstone and Dhrystone benchmarks. Another complicating factor is the environment in which the benchmark is run. These could include operating system version, compiler version, memory speed, I/O devices, etc. Unless these are spelled out in detail it is difficult to interpret the results of a standard benchmark.

Three new organizations have been formed recently with the goal of providing more meaningful benchmarks for comparing the capability of computer systems for doing different types of work. The Transaction Processing Performance Council (TPC) was founded in 1988 at the initiative of Omri Serlin to develop online teleprocessing (OLTP) benchmarks. Just as the TPC was organized to develop benchmarks for OLTP the Standard Performance Evaluation Corporation (SPEC) is a nonprofit corporation formed to establish, maintain, and endorse a standardized set of benchmarks that can be applied to the newest generation of high-performance computers and to assure that these benchmarks are consistent and available to manufacturers and users of high-performance systems. The four founding members of SPEC were Apollo Computer, Hewlett-Packard, MIPS Computer Systems, and Sun Microsystems. The Business Applications Performance Corporation (BAPCo) was formed in May 1991. It is a nonprofit corporation that was founded to create for the personal computer user objective performance benchmarks that are representative of the typical business environment. Members of BAPCo include Advanced Micro Devices Inc., Digital Equipment, Dell Computer, Hewlett-Packard, IBM, Intel, Microsoft, and Ziff-Davis Labs.

6.6.1 The Standard Performance Evaluation Corporation (SPEC)

In October 1989 the Standard Performance Evaluation Corporation (SPEC) released its first set of 10 benchmark programs known as Release 1.0. The SPEC Suite Release 1.0 consists of 10 CPU-intensive benchmarks derived from or taken directly from applications in the scientific and engineering disciplines. Results are given as performance relative to a VAX 11/780 using VMS compilers. Thus, if t_i is the wall clock time to perform benchmark *i* on the test machine and t_{vaxi} is the wall clock time to run the benchmark on a VAX 11/780, then the result for benchmark *i* is computed as $r_i = t_{\text{vaxi}} / t_i$. The final unit is the SPECmark, which is the geometric mean of the individual benchmarks. Thus it is $(r_1 \times r_2 \times \ldots \times r_{10})^{1/10}$ where r_i is the result from benchmark *i*.

On January 15, 1992 SPEC announced the availability of two new benchmark suites. They are the CPU-intensive integer benchmark suite (CINT92) and the CPU-intensive floating-point Suite (CFP92).

The new integer suite consists of six new benchmarks which represent application areas in circuit theory, LISP interpreter, logic design, text compression algorithms, spreadsheet, and software development.

The new floating-point suite is comprised of 14 benchmarks, 5 of which are single precision, representing application areas in circuit design, Monte-Carlo simulation, quantum chemistry, optics, robotics, quantum physics, astrophysics, weather prediction, and other scientific and engineering problems.

Table 6.1. SPEC Benchmark Results

Model	HP/ 705	HP/ 750	IBM/ 220	IBM/ 970	Sun SS2
SPECmark rel 1.0	34.6	86.5	25.9	100.3	25.0
SPECint92 rel 2.0	21.9	51.1	15.9	47.1	21.8
SPECfp92 rel 2.0	33.0	84.9F	22.9	93.6	22.8
Rated MIPS (VAX)	40.0	76.0	n/a	n/a	28.0
Rated MFlops (dbl)	8.4	23.7	6.5	n/a	4.2
CPU Clock MHz	35	66	33	50	40

Table 6.2. More SPEC Benchmark Results

Model	Intel Pentium	HP 735	DEC Alpha	IBM RS/ 6000	Sun SS
SPECint92 rel 2.0	64.5	80.0	65.3	59.2	52.6
SPECfp92 rel 2.0	56.9	150.6	111.0	124.8	64.7
CPU Clock MHz	66	99	135	62.52	40/

Rather than have one composite number for the combined two benchmark suites SPEC provides a separate metric for CINT92 and for CFP92. SPECint92 is the composite metric for CINT92. It is the geometric mean of the SPECratios of the six integer benchmarks. The SPECratio for a benchmark on a given system is the quotient derived by dividing the SPEC Reference Time for that benchmark (run time on a DEC VAX 11/780) by the run time for the same benchmark on that particular system. SPECfp92 is the composite metric for CFP92 and is the geometric mean of the SPECratios of the fourteen floating-point benchmarks. We provide some representative SPEC benchmark results in Tables 6.1 and 6.2. These results are those reported to SPEC by the manufacturers. Note that IBM no longer reports MIPS results.

In Table 6.1 HP/705 is shorthand for Hewlett-Packard HP 9000 Series 705 and similarly for HP/750. IBM/220 is shorthand for IBM RS/6000 Model 220 and similarly for IBM/970. Sun SS2 is an abbreviation for Sun SPARCstation 2. In Table 6.2 SS is shorthand for SuperSPARC. All the results in Table 6.2 were reported in [Boudette 1993]. In his article Boudette also included the performance results reported by Intel for the Intel 66 MHz Pentium, the 60 MHz Pentium, the 33/66 MHz 486DX2, the 50 MHz 486DX, the 25/50 MHz 486DX2, the 33 MHz 486DX, and the 25 MHz 486DX based on the internal Intel benchmark Icomp. These benchmark results indicate that the 66 MHz Pentium almost doubles the performance of the 33/66 MHz 486DX2 which is 78.9 percent faster than the 33 MHz 486DX.

In addition to reporting the composite metrics SPECint92 and SPECfp92 manufacturers report the performance on each individual benchmark. This helps users better position different computers relative to the work to be done. The floating-point suite is recommended for comparing the floating-point-intensive (typically engineering and scientific applications) environment. The integer suite is recommended for environments that are not floating-point-intensive. It is a good indicator of performance in a commercial environment. CPU performance is one of the indicators of commercial environment performance. Other components include disk and terminal subsystems, memory, and OS services. SPEC has announced that benchmarks are being readied to measure overall throughput, networking, and disk input/output for release in 1992 and 1993. Currently SPEC benchmarks run only under UNIX.

6.6.2 The Transaction Processing Performance Council (TPC)

The Transaction Processing Performance Council (TPC) is made up of a number of member companies representing a wide spectrum of the computer industry. Members include big U. S. vendors such as Hewlett-Packard, IBM, Digital Equipment, and Amdahl, foreign computer companies such as NEC, Fujitsu, Hitachi, and Bull, as well as major database software vendors such as Computer Associates and Oracle.

TPC publishes benchmark specifications that regulate the running and reporting of transaction processing performance data. It is the goal of each specification to provide a "level playing field" so that customers are able to make objective comparisons among performance data published by competing vendors on different system platforms. Before a hardware or software vendor can claim performance figures with a TPC benchmark the vendor must file a Full Disclosure Report (FDR) with the TPC explaining exactly how the benchmark was performed. While it is not a formal requirement, vendors reporting TPC numbers are strongly urged to employ an outside auditor to certify the performance claims.

Each FDR must be on file with the TPC administrator's office for the claimed TPC results to be valid. Once the FDR is filed with the administrator, it receives a "submitted for review" status. Copies of the FDR are circulated to all members of the TPC who then have 60 days to review and challenge, the report on the basis that it is not in conformance with the TPC benchmark specifications. Questions and challenges are initially submitted to the TPC's Technical Advisory Board (TAB), which reviews the issue and provides the TPC Council with a recommendation. If an FDR is challenged, the council must decide whether the FDR is compliant or not within a period of 60 days. If there is no challenge or Council ruling of non-compliance within this 60 day review period, the FDR passes into "accepted" status.

One of the first tasks the TPC set for itself was to provide a formal definition of the de facto standard Debit-Credit benchmark and its derivative TP1. The only public definition of the Debit-Credit benchmark was a loosely defined benchmark described in [Anon *et al.* 1985]. Vendors who published Debit-Credit numbers tended to take liberties with the definition in order to make their systems look good.

In November 1989, the TPC formally published its first benchmark specification, TPC Benchmark A (TPC-A), with a workload that bears some resemblance to Debit-Credit. TPC-A is a complete system benchmark and simulates an environment in which multiple users, using terminals, are accessing and updating a common database over a local or wide-area network (thus, the terms "tpsA-local" and "tpsA-wide"). The TPA-A benchmark uses the human and computer operations involved in a typical banking automated teller machine (ATM) transaction as a simplified model to represent a wide array of OLTP business transactions. Results of the benchmark are expressed in TPS (transactions per second) and in $/TPS or dollars per TPS.[At first it was planned to represent the cost in units of thousands of dollars per TPS ($K/TPS) but it was found to be too complicated for business executives to think in those terms.] The TPS rating is equal to the number of transactions completed per unit of time provided that 90 percent of the transactions have a response time of two seconds or less. The $/TPS is the total cost of the system tested divided by the obtained TPS rating. This is intended as a price-performance measure so the lower the result, the better the performance. The total system cost includes all major hardware and software components (including terminals, disk drives, operating system and database software as required by benchmark specifications), support, and 5 years of maintenance costs.

The second TPC benchmark, called TPC Benchmark B (TPC-B), is intended as a replacement for TP1. TPC-B was approved in August 1990 and is primarily a database server test in which streams of transactions are submitted to a database host/server in a batch mode. The database operations associated with TPC-B transactions are similar to those of TPC-A, but there are no terminals or end-users associated with the TPC-B benchmark. Results of this benchmark are the same as those for the TPC-A benchmark: TPS and $/TPS.

In Table 6.3 we present some of the results reported by the TPC on March 15–16, 1992. The TPC-A results are local results.

Although the TPC-A and TPC-B benchmarks have been widely accepted there has been some criticism of some features of these benchmarks. The most severe charge against the two benchmarks is that neither truly represents any actual segment of the commercial computing marketplace. Another complaint is that the TPS rating is too sensitive to the requirement that 90 percent of all transactions must have a response time not exceeding 2 seconds. The TPC-A benchmark has been criticized for being a single-transaction workload although most commercial workloads have a batch component. The TPC-B benchmark has a batch but no online component. To answer these complaints the TPC has developed a new benchmark called TPC-C that is considered to be an order-entry benchmark.

Table 6.3. Representative TPC-A and TPC-B Results

Computer	TPS-A	$/TPS	TPS-B	$/TPS-B
DECsystem 5100	10.60	22,774	28.20	2,345
DECsystem 5500	21.10	18,101	40.60	3,944
HP 9000 Series 817S	51.27	11,428	64.79	1,940
HP 9000 Series 842S	33.00	25.500	81.10	2,900
IBM AS/400 Model D10	6.50	17,850		
IBM RS/6000 Model 320			31.40	2,806
Sun SPARCserver 690 MP	95.41	8,854	134.90	2.764

TPC-C simulates an order-entry application with a number of transaction types. These transactions include entering and delivering orders, recording payments, checking the status of orders, and monitoring the level of stock at the warehouses.

The most frequent transaction consists of entering a new order which, on the average, consists of 10 different items. Each warehouse maintains stock for the 100,000 items in the catalog and fills orders from that stock. Since one warehouse will often not have all 10 of the items ordered in stock, TPC-C requires that about 10 percent of all orders must be supplied by another warehouse. Another frequent transaction is the recording of a payment received from a customer. Less frequent transactions include operator request of the status of a previously placed order, processing a batch of 10 orders for delivery, or querying the system for the level of stock at the local warehouse. The performance metrics reported by TPC-C are tpm-C, the average number of orders processed per minute, and $/tpm, the cost per tpm-C. The latter is calculated in the same way that the cost per TPS is calculated for TPC-A and TPC-B. The TPC-C benchmark was approved by the council in July 1992.

The TPC-A and TPC-B benchmarks are not directly usable for making purchase decisions because neither of them can be matched with an actual application. However, they do provide information to those who are planning to develop OLTP applications. By reading TPC-A and TPC-B reports from different vendors application developers can obtain rough ideas about the performance of competing computer systems as well as relative costs. However, developers who have applications similar to that described by the TPC-C benchmark are able to make at least a rough estimate of what model of computer is needed if they read the FDRs in detail for the machines of interest.

Table 6.4. Representative TPC-C Results

Computer	tpsC	$/tpsC	Software
HP 3000/947	105.26	$4,171	Allbase/SQL VF0.23
HP 3000/957	180.24	$3,225	Allbase/SQL VF0.23
IBM AS/400 9404 E10	33.81	$2,462	OS/400 Int. Rel. DB V2 Rel 2
IBM AS/400 9404 E35	54.14	$3,483	OS/400 Int. Rel. DB V2 Rel 2

The TPC-C results reported in Table 6.4 are from [Boudette 1993].

6.6.3 Business Applications Performance Corporation

The Business Applications Performance Corporation (BAPCo) benchmarks are intended to provide a means of comparing the performance of industry standard architecture systems while using commercially available applications. The BAPCo benchmarks are designed to measure hardware performance, not software. Three workloads are planned: stand-alone, multitasking and network. The stand-alone workload uses DOS and Windows applications and is the first product of BAPCo. The availability of the SYSmark92 benchmark suite was announced on May 27, 1992. It is the first stand-alone benchmark. It measures the personal computer's speed in word processing, spreadsheet, database, desktop graphics, desktop publishing, and software development. The application selections for this release are as follows:

Word Processing
MS Word for Windows 1.1
Wordperfect 5.1
Spreadsheet
Lotus 123 R 3.1+
Excel 3.0
Quattro Pro 3.0
DataBase
dBASE IV 1.1
Paradox 3.5
Desktop Graphics
Harvard Graphics 3.0
Desktop Publishing
Pagemaker 4.0
Software Development
Borland C++ 2.0
Microsoft C 6.0

The metric used to quantify performance is scripts per minute. This metric is calculated for each application and then combined to yield a performance metric for each category. Thus there is a metric for word processing, spreadsheets, database, desktop graphics, desktop publishing, and software development. According to Strehlo [Strehlo 1992], the scoring is calibrated so that a typical 33 MHz 486 computer will score approximately 100. One could use the output from the SYSmark92 benchmark performed on a number of different personal computers to help decide what personal computers to buy for people who have similar workloads. For example, for users in a group that makes a lot of spreadsheet calculations, the spreadsheet rating can be used to compare the usefulness of different personal computers for making spreadsheet computations. Then all the PCs that satisfy your spreadsheet rating criterion can be analyzed relative to other factors such as price, ease-of-use, quality, support policies, training requirements, if any, etc., to make the final purchase decision. Part of any decision should involve allowing some of the final users to test the machines to see which ones they like.

6.6.4 Drivers (RTEs)

To perform some of the benchmarks we have mentioned, such as the TPC benchmarks TPC-A and TPC-C, a special form of simulator called a *driver* or *remote terminal emulator* (RTE) is used to generate the online component of the workload. The driver simulates the work of the people at the terminals or workstations connected to the system as well as the communication equipment and the actual input requests to the computer system under test (SUT in benchmarking terminology). An RTE, as shown in Figure 6.1, consists of a separate computer with special software that accepts configuration information and executes job scripts to represent the users and thus generate the traffic to the SUT. There are communication lines to connect the driver to the SUT. To the SUT the input is exactly the same as if real users were submitting work from their terminals. The benchmark program and the support software such as compilers or database management software are loaded into the SUT and driver scripts representing the users are placed on the RTE system. The RTE software reads the scripts, generates requests for service, transmits the requests over the communication lines to the benchmark on the SUT, waits for and times the responses from the benchmark program, and logs the functional and performance information. Most drivers also have software for recording a great deal of statistical performance information.

Most RTEs have two powerful software features for dealing with scripts. The first is the ability to capture scripts from work as it is being performed. The second is the ability to generate scripts by writing them out in the format understood by the software. An example of the first kind of script is given in Table 6.5. This script was automatically generated by Helen Fong, by using the collector facility of Wrangler, the driver we use at the Hewlett-Packard Performance Technology Center. As she performed the described operations at her workstation, the collector recorded what she did. She added comments to explain what she was doing and streamlined the scripts. Comments can be identified because they start with "!*." When Helen was through she asked the reduction program to generate the script shown. Once scripts are available they can be combined to form a terminal workload. Thus 25 copies of the script in Table 6.3 can be generated and combined with other scripts from other online work to form a terminal workload class. This workload is then executed on the SUT by the RTE.

Figure 6.1. Remote Terminal Emulator (RTE)

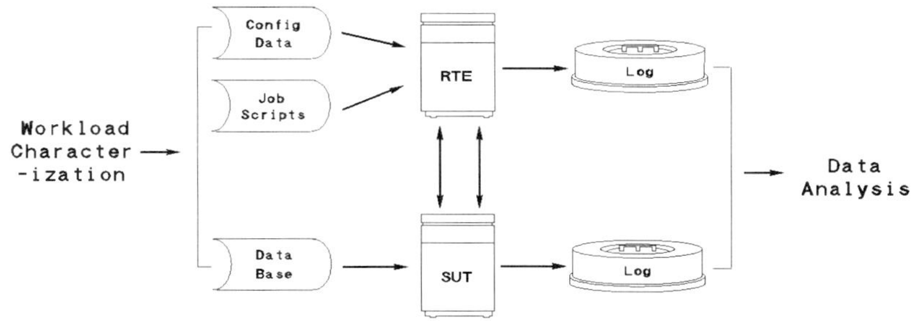

Table 6.5. A Wrangler Script

```
!SCRIPT  AUTOCAPTURE
!*
!*     Automated MPE V/E Script For Ldev 120
$CONTROL ERRORS=10, WARN
!*
!* Set the terminal line transmission speed to 960,
emulation
!* mode to character mode
!*
!SET speed=960, mode=char, type=0
!SET eor=nul
!*TIMER = 15:32:44
!LOGON
!* Generate a message to the SUT to logon and wait 70
decisec-
!* onds from the receipt of a PROMPT character from the
SUT
!* before sending the next message.
!*
!SEND "hello manager.sys", CR
!WAIT 0,      70
!*
!*  Generate a message to the SUT to execute GLANCE
!*
!SEND "run glancev.pub.sys", CR
```

```
!WAIT 0,        3
!*
!*  Generate a message to the SUT to examine the GLOBAL
screen
!*
!SEND "g"
!WAIT 0,        0
!*
!*  Generate a message to the SUT to EXIT from GLANCE
!*
!SEND "e"
!WAIT 0,       26
!*  Generate a message to the SUT to logoff the MPE
session
!*
!SEND "BYE", CR
!*TIMER = 15:33:22
!LOGON
!* End Of Script
!*TIMER = 15:33:23
!END
```

All computer vendors have drivers for controlling their benchmarks. Since there are more IBM installations than any other kind, the IBM Teleprocessing Network Simulator (program number 5662-262, usually called TPNS) is probably the best known driver in use. TPNS generates actual messages in the IBM Communications Controller and sends them over physical communication lines (one for each line that TPNS is emulating) to the computer system under test.

TPNS consists of two software components, one of which runs in the IBM mainframe or plug compatible used for controlling the benchmark and one that runs in the IBM Communications Controller. TPNS can simulate a specified network of terminals and their associated messages, with the capability of altering network conditions and loads during the run. It enables user programs to operate as they would under actual conditions, since TPNS does not simulate or affect any functions of the host system(s) being tested. Thus it (and most other similar drivers including WRANGLER, the driver used at the Hewlett-Packard Performance Technology Center) can be used to model system performance, evaluate communication network design, and test new application programs. A driver may be much less difficult to use than the development of some detailed simulation models but is expensive in terms of the hardware required. One of its most important uses is testing new or modified online programs both for accuracy and performance. Drivers such as TPNS or WRANGLER make it possible to utilize all seven of the uses of benchmarks described by Artis and Domanski. Kube in [Kube 1981] describes how TPNS has been used for all these activities. Of course the same claim can be made for most commercial drivers.

6.6.5 Developing Your Own Benchmark For Capacity Planning

Unless your objectives are very limited or your workload is very simple, developing your own benchmark for predicting future performance on your current system or an upgraded system is rather daunting. By "predicting future performance" we mean predicting performance with the workload you forecast for the future. Experienced benchmark developers complain about "the R word," that is, developing a benchmark that is truly representative of your actual or future workload. You may be thinking, "Yes, but if my computer system has a terminal workload with no batch classes, then I can use an RTE to capture the scripts from my actual workload. Then all I have to do to run a benchmark is to run these scripts from the RTE suitably amplified to account for growth." However, even in this simple, unusual case, it requires major resources and skills to run representative benchmarks. Recall that an RTE runs on a separate computer system from the SUT (system under test) and often runs on a more powerful computer than the SUT. This is expensive because it also must have all the hardware required to deliver the simulated requests for service to the SUT. During the hours or days that the benchmark is run neither the RTE computer nor the SUT computer can be used for doing useful work. Recall, also that the RTE is a simulator (emulation is a form of simulation) so you have the usual problems with starting up a simulation run. Just as with all simulation runs, such runs do not generate useful information until the system has reached the steady-state. The problem is in determining when the steady-state has been reached. Assuming you are successful in determining when the steady state is reached and thus can ignore the performance data that occurred before that time, there are difficulties in interpreting the results of the benchmark runs; I say runs because you must make multiple runs to ensure that the benchmark is repeatable. There are a number of entities that you would like to determine from a benchmark study that sound very simple but that, in practice, are nearly impossible to accomplish. For example, suppose you are currently supporting 40 active users during the peak period of the day. You would like to validate your benchmark by running it first with 40 active users, measure the performance and check the measured benchmark performance against the measured performance of the real system with 40 active people at their terminals. This is very difficult to do because of the difficulty of getting the RTE to generate the exact load on the system that the 40 users would even though it is using the captured scripts from the 40 users. You can't just issue a command to the RTE to emulate 40 users. You must experiment with different think times until the load generated on the system is close to the load generated by 40 real people. What can be achieved by benchmarking with the RTE is to find the maximum load that your current system will support at a performance level that is acceptable to the users. Then you will have the challenge and the expense of obtaining time on a more powerful or several more powerful computer systems that you want to consider for upgrade options. Most installations that decide to do their own benchmarks must depend upon using the facilities of their computer vendor. Most large computer companies have benchmarking facilities that are available to their customers for a price. Most are also prepared to provide experienced people with benchmarking experience to help with the benchmarking process.

Since very few computer systems run with only terminal workload classes, most benchmarking experts recommend that you include one or more batch workload classes in your benchmark. See, for example, the chapter by Tom Sawyer (yes, there really is a Tom Sawyer!) in [Gray 1991]. The title of the chapter is "Doing Your Own Benchmark." Sawyer says:

> Batch work should be included in the benchmark. Most shops that run online work during the day discover that the batch window is a critical resource. If no batch work is included in your measurements, the vendors may be tempted to use devices that have good online characteristics but have weak batch performance. For instance, disk drives connected using the SCSI interface perform well in online operations but do not have the sequential capabilities of IPI drives.
>
> We shall assume that the goals of the proposed system include the ability to run batch jobs without degrading the performance of the online work.
>
> You may also want to consider benchmarking a few key batch jobs that must be run frequently and can be run when the online environment need not be up.

If, like most installations, batch jobs are run on your computer system with your online (terminal) workload classes, you can use your RTE to capture the scripts in which batch jobs are launched. However, it can be a real challenge to construct a representative batch workload if you run a number of different batch jobs with very different resources requirements. The benchmark section of [Howard] describes the rather tedious procedure for construction a representative batch workload.

In spite of all the difficulties and challenges I have cited, it is possible to construct representative and useful benchmarks. Computer manufacturers couldn't live without them and some large computer installations depend upon them. However, constructing a good benchmark for your installation is not and easy task and is not recommended for most installations. In their excellent paper [Dongarra, Martin, and Worlton 1987], Dongarra *et al.* warned:

> Evaluators who do benchmarks in pursuit of a single, all-encompassing number can end up with meaningless results if they commit these errors:
>
> 1. Neglecting to characterize the workload.
> 2. Selecting kernels that are too simplistic.Using programs or algorithms adapted to a specific computer system.
> 3. Running benchmarks under inconsistent conditions.
> 4. Selecting inappropriate workload measures.
> 5. Neglecting the special needs of the users.
> 6. Ignoring the difference between the frequency and the duration of execution.

Note that the authors define a kernel to be the central portion of a program, containing the bulk of its calculations, which consumes the most execution time.

Clark, in his interesting paper [Clark 1991], provides a report on his experiences at his installation in developing and running their first benchmark. Their benchmark was what Clark calls a proof-of-concept (POC)" benchmark. Clark describes this type of benchmark as follows:

> This project embraces both the hardware *and* software aspects of a computer system. It has a specific purpose for the company; establishing reasonable evidence that it is possible to process a workload on a conceived architectural platform, operating system, or network. Accuracy, while always wel-

> comed *and* encouraged, is not the primary consideration, and wider bounds of accuracy are permitted providing they are stated and understood. Expedience will have a high priority, as dictated by management deadlines.

Clark does not reveal the exact purpose of the benchmark study. However, it appears that the feasibility of moving an application that was running under CICS on an IBM mainframe to an open platform was to be determined. On the open platform SQL would be used to access the data. For the latter part of the benchmark it was necessary to simulate SQL transactions using a relational database management system. Clark discusses the planning, team involvement, establishing control over the vendor benchmark personnel, scope, workload, data, driving the benchmark, documentation, and the final report. Clark is an experienced performance analyst and had access to advice from Bernard Domanski, an experienced benchmarker, so his chances for success were greatly enhanced over that to be expected for someone relatively new to computer performance analysis. For Clark's study workload characterization and the generation of test data were especially challenging.

Exercise 6.4

You are the lead performance analyst at Information Overload. You have excellent rapport with your users who provide very good feedback on their workload growth so that you can accurately predict the demands on your computer system. Your performance studies show that your current computer system will be able to support your workload at the level required by the service level agreement you have with your users for only six more months. You have prepared a list of three different computer systems from three different vendors that you feel are good upgrade candidates based on your modeling of the three systems. Clarence Clod, the manager of your installation, insists that you must conduct benchmark studies on the three different computer systems using a representative benchmark that you must develop before a new system can be ordered. Your biggest challenge in complying with his orders will be:

(a) Constructing a truly representative benchmark in time to run it on the three systems.

(b) Assuming that you succeed with (a), running the benchmark successfully on the three candidate systems.

(c) Assuming you succeed with (a) and (b), analyzing the results of the three studies in a way that will give you great confidence that you can make the correct choice.

(d) None of the above.

Exercise 6.5

You are the manager of a group of engineers who are using a simulation package on their workstations to design electronic circuits. The simulation package is heavily dependent upon floating-point calculations. The engineers complain that their projects are getting behind schedule because their workstations are so slow. You obtain authorization from your management to replace all your workstations. As you read the literature from different vendors on their workstations what benchmarks or performance metrics will be of most importance to you?

6.7 Solutions

Solution to Exercise 6.1

The two runs requested follow. They were made on my Hewlett-Packard workstation and thus took less time but yielded exactly the results made on my home 33MHz 486DX IBM PC compatible.

```
In[4]:= simmm1[0.9, 1.0, 11, 1500, 500]//Timing
The mean value of time in system at end of warmup is
6.1455
Mean time in system is 9.86683
95 percent confidence interval is
8.77123 to 10.9624

Out[4]= {179.42 Second, Null}

In[5]:= simmm1[0.9, 1.0, 11, 1500, 2000]//Timing
The mean value of time in system at end of warmup is
6.1455
Mean time in system is 9.85506
95 percent confidence interval is
9.12232 to 10.5878

Out[5]= {709.79 Second, Null}
```

The exact value of the average steady-state response time for an M/M/1 queueing system with server utilization 0.9 is 10. For the first run the estimate of this quantity is 9.86683, the 95 percent confidence interval contains the correct value, and the length of the confidence interval is 2.19117. For the second run the estimated value of the average response time is 9.85506 (not quite as good an estimate as we obtained for the shorter first run), the confidence interval contains the correct value, and the length of the confidence interval is 1.46548.

Solution to Exercise 6.2

The output from the following runs of **ran** show periods of the values of *a* not previously considered.

```
In[5]:= m =13

Out[5]= 13

In[6]:= seed =2

Out[6]= 2

In[7]:= n = 13

Out[7]= 13

In[8]:= ran[2, m, n, seed]

Out[8]= {2, 4, 8, 3, 6, 12, 11, 9, 5, 10, 7, 1, 2}

In[9]:= ran[3, m, n, seed]

Out[9]= {2, 6, 5, 2, 6, 5, 2, 6, 5, 2, 6, 5, 2}

In[10]:= ran[4, m, n, seed]

Out[10]= {2, 8, 6, 11, 5, 7, 2, 8, 6, 11, 5, 7, 2}

In[11]:= ran[5, m, n, seed]

Out[11]= {2, 10, 11, 3, 2, 10, 11, 3, 2, 10, 11, 3, 2}
```

```
In[12]:= ran[8, m, n, seed]

Out[12]= {2, 3, 11, 10, 2, 3, 11, 10, 2, 3, 11, 10, 2}

In[13]:= ran[9, m, n, seed]

Out[13]= {2, 5, 6, 2, 5, 6, 2, 5, 6, 2, 5, 6, 2}

In[14]:= ran[10, m, n, seed]

Out[14]= {2, 7, 5, 11, 6, 8, 2, 7, 5, 11, 6, 8, 2}

In[15]:= ran[11, m, n, seed]

Out[15]= {2, 9, 8, 10, 6, 1, 11, 4, 5, 3, 7, 12, 2}

In[16]:= ran[12, m, n, seed]

Out[16]= {2, 11, 2, 11, 2, 11, 2, 11, 2, 11, 2, 11, 2}

In[4]:= ran[7, 13, 13, 2]

Out[4]= {2, 1, 7, 10, 5, 9, 11, 12, 6, 3, 8, 4, 2}
```

From the above runs of **ran** and the runs performed earlier we construct Table 6.6.

Table 6.6. Results From Exercise 6.2

Multi plier	Period	Multi plier	Period	Multip lier	Period
2	12	3	3	4	6
5	4	6	12	7	12
8	4	9	3	10	6
11	12	12	12		

It is interesting to note that there are four full period multipliers—2, 6, 11, 12.

Solution to Exercise 6.3

```
In[3]:= <<work.m

In[5]:= SeedRandom[47]

In[6]:= y = Table[Random[ExponentialDistribution[1/
10]], {5000}];

In[7]:= chisquare[0.02, y, 10]

"p is "0.1262315175895422
"q is "0.873768482410458
"The sequence passes the test."
```

This solution was made using Version 2.1 of *Mathematica*; Version 2.0 yields slightly different values for p and q because the output of SeedRanom[47] is different for the two versions of *Mathematica*.

Solution to Exercise 6.4

One could make a good case for any of the answers. Benchmark experts such as Professor Domenico Ferrari at UC Berkeley claim that the most difficult part of a benchmarking study is constructing a representative benchmark. If there are no batch workload classes, that is, all workload classes are terminal class workloads, it may be possible to capture a representative workload using a remote terminal emulator. This may be more difficult if, rather than dumb terminals, the users are using workstations to access the computer system.

Even if you have constructed a representative benchmark, running the benchmark properly requires some expertise that comes only with experience. This is not as daunting if the workload consists only of terminal classes and the benchmark is run with a sophisticated remote terminal emulator such as TPNS or Wrangler.

Properly interpreting the results of simulation runs is anything but straightforward, too, so one could make a case for this being the most difficult problem.

Finally, none of the above would be a good choice for you if your financial as well as personnel resources are limited. If you have a big budget or the systems you are considering are very expensive, you can probably persuade the vendors to run the benchmarks for you by their experienced benchmark personnel. You must keep in mind, however, that each vendor will certainly be highly motivated to try to convince you that their system is the most effective for your workload.

Solution to Exercise 6.5

The benchmark from SPEC that should be of interest to you is the benchmark in the new floating-point suite representing areas in circuit design. You can compare the SPECratio of this individual benchmark for different workstations. You will probably be interested in the composite floating-point suite metric SPECfp92 as well. Another important consideration is how easy it will be to port your simulation program to your new workstations.

6.8 References

1. Anon *et al.*, "A measure of transaction processing power," *Datamation*, April 1, 1985, 112—118.

2. H. Pat Artis and Bernard Domanski, *Benchmarking MVS Systems*, notes from the course taught January 11—14 1988 at Tyson Corner, VA.

3. Jon Bentley, "Some random thoughts," *Unix Review*, June 1992, 71—77.

4. Neal Boudette, "Intel gears Pentium to drive continued 486 system sales," *PCWEEK*, February 15, 1993.

5. Paul Bratley, Bennett L. Fox, and Linus E. Schrage, *A Guide to Simulation,* Second Edition, Springer-Verlag, New York, 1987.

6. James D. Calaway, "SNAP/SHOT VS BEST /1," *Technical Support*, March 1991, 18—22.

7. Philip I. Clark, "What do you really expect from a benchmark?: a beginners' perspective," *CMG '91 Proceedings*, Computer Measurement Group, 826—832.

8. Jack Dongarra, Joanne L. Martin, and Jack Worlton, "Computer benchmarking: paths and pitfalls," *IEEE Spectrum*, July 1987, 38—43.

9. Tony Engberg, "Performance: questions worth asking," *Interact*, August 1988, 50—61.

10. Paul J. Fortier and George R. Desrochers, *Modeling and Analysis of Local Area Networks*, CRC Press, Boca Raton, FL, 1990.

11. Martin Gardner, *Mathematical Carnival*, Mathematical Association of America, Washington, DC, 1989.

12. Martin Gardner, *Fractal Music, Hypercards and More ...*, W. H. Freeman, New York, 1992.

13. J C. Gibson, *IBM Technical Report TR-00.2043*, June 18, 1970.

14. Jim Gray, Ed, *The Benchmark Handbook*, Morgan Kaufmann Publishers, San Mateo, CA, 1991.

15. Richard W. Hamming, *The Art of Probability for Scientists and Engineers*, Addison-Wesley, Reading, MA, 1991.

16. Phillip C. Howard, *Capacity Management Handbook Series, Volume 1: Capacity Planning*, Institute for Computer Capacity Management, Phoenix, AZ, 1990.

17. Leonard Kleinrock, *Queueing Systems, Volume 1: Theory*, John Wiley, New York, 1975.

18. Donald E. Knuth, T*he Art of Computer Programming: Seminumerical Algorithms* Second Edition, Addison-Wesley, 1981.

19. Hisashi Kobayashi, *Modeling and Analysis: An Introduction to System Performance Evaluation Methodology*, Addison-Wesley, Reading, MA, 1978.

20. C. B. Kube, *TPNS: A Systems Test Tool to Improve Service Levels*, IBM Washington Systems Center, GG22-9243-00, 1981.

21. Stephen S. Lavenberg, Editor, *Computer Performance Modeling Handbook,* Academic Press, New York, 1983.

22. M. H. MacDougall, *Simulating Computer Systems:Techniques and Tools*, The MIT Press, Cambridge, MA, 1987.

23. Edward A, MacNair and Charles H. Sauer, *Elements of Practical Performance Modeling,* Prentice-Hall, Englewood Cliffs, NJ, 1985.

24. George Marsaglia, "Random numbers fall mainly in the plains," *Proceedings of the National Academy of Sciences*, 61, 25—28.

25. George Marsaglia and Arif Zaman, "A new class of random number generators," *The Annals of Applied Probability*, 1(3), 1991, 462—480.

26. George Marsaglia "A current view of random number generators," *Computer Science and Statistics: 16th Symposium on the Interface*, Elsevier, New York, 1985, 1—8.

27. George Marsaglia and Arif Zaman, "The random number generator ULTRA," Draft of Research Report, Department of Statistics and Supercomputer Computations Research Institute, The Florida State University, 1992.

28. Byron J. T. Morgan, *Elements of Simulation*, Chapman and Hall, London, 1984.

29. Stephen Morse, "Benchmarking the benchmarks," Network Computing, February 1993, 78—84.

30. Stephen K. Park and Keith W. Miller, "Random number generators: good ones are hard to find," *Communications of The ACM,* October 1988, 1192—1201.

31. Ivars Peterson, "Monte Carlo physics: a cautionary lesson," *Science News*, December 19 & 26, 1992, 422.

32. Robert Pool, "Computing in science," *Science*, April 3, 1992, 44–62.

33. Rand Corporation, *A Million Random Digits With 100,000 Normal Deviates*, The Free Press, Glencoe, IL 1955.

34. Omri Serlin, "MIPS, Dhrystones and other tales," *Datamation*, June 1986, 112—118.

35. Kevin Strehlo, "BAPCo benchmark offers worthy performance test," *InfoWorld*, June 8, 1992.

36. Reinhold P. Weicker, "An Overview of Common Benchmarks," *IEEE Computer*, December 1990, 65—75.

37. Peter D. Welch, "The statistical analysis of simulation results," in *Computer Performance Modeling Handbook*, Stephen S. Lavenberg Ed., Academic Press, New York, 1983.

Chapter 7 Forecasting

I know of no way of judging the future but by the past.
Patrick Henry

7.1 Introduction

As Patrick Henry suggests, forecasting means predicting the future from the past. In ancient times this was done by examining chicken entrails or consulting an oracle. In modern times the concept of time series analysis has developed to help us predict the future. Forecasting is most useful in predicting workload growth but may sometimes be used to predict CPU utilization or even response time. Forecasting using time series analysis is essentially a form of pattern recognition or curve fitting. The most popular pattern is a straight line but other patterns sometimes used include exponential curves and the S-curve. One of the keys to good forecasting is good data and the source of much useful data is the user community. That is why one of the most popular and successful forecasting techniques for computer systems is forecasting using natural forecasting units (NFUs), also known as business units (BUs) and as key volume indicators (KVI). The users can forecast the growth of natural forecasting units such as new checking accounts, new home equity loans, or new life insurance policies sold much more accurately than computer capacity planners in the installation can predict future computer resource requirements from past requirements. If the capacity planners can associate the computer resource usage with the natural forecasting units, future computer resource requirements can be predicted. For example, it may be true that the CPU utilization for a computer system is strongly correlated with the number of new life insurance policies sold. Then, from the predictions of the growth of policies sold, the capacity planning group can predict when the CPU utilization will exceed the threshold which will require an upgrade.

7.2 NFU Time Series Forecasting

NFU forecasting is a form of time series forecasting. However, a number of aspects of time series forecasting need to be discussed before discussing NFU forecasting. Time series forecasting is a discipline that has been used for applications such as studying the stock market, the economic performance of a nation, population trends, rainfall, and many others. An example of a time series that we might study as a computer performance analyst is $u_1, u_2, u_3, \ldots, u_n, \ldots$ where u_i is the maximum CPU utilization on day i for a particular computer system.

All the major statistical analysis systems such as SAS and Minitab provide tools for the often complex calculations that go with time series analysis. For the convenience of computer performance analysts who have Hewlett-Packard computer equipment the Performance Technology Center has developed HP RXForecast for HP 3000 MPE/iX computer systems and for HP 9000 HP-UX computer systems. We discuss how RXForecast can be used for business unit (NFU) forecasting in the next section.

Several concepts are important in studying time series. The first is the *trend*, which is the most important component of a time series. Trend tells us whether the values in the series are increasing or decreasing in the long run. What "long run" means for a specific case is sometimes difficult to determine. Series that neither increase nor decrease are called *stationary*. Chatfield [Chatfield 1984] defines trend as "long term change in the mean." For time series with an increasing or decreasing trend, the only kind of interest to us, we are also interested in the pattern of the trend. The most common patterns for computer performance data are linear, exponential, and S-curve shaped.

A basic problem in time series analysis is separating the trend from three other components that tend to mask the trend. The first of these components is *seasonal variation* or *seasonality*. A seasonal pattern has a constant length and occurs again and again on a regular basis. Thus a toy company with most of its sales occurring at Christmas time could expect an annual seasonality in its computer workload, as would a firm that prepares income tax returns. Companies that have a weekly basis for reporting may have a weekly seasonality, those with a monthly reporting structure a monthly seasonality, etc.

Some time series have a cyclical pattern that is usually oscillatory and has a long period. For example, some economists believe that economic data are driven by business cycles with a period varying between 5 and 7 years. This cycle could have an effect on computer usage. There may be other cyclic patterns in computer performance data as well. If so, it is very useful to know about such cycles.

There often is a *random* component to time series values. By this we mean an unpredictable component due to random effects.

Statisticians have devised methods that allow one to detect and remove the seasonal component if one exists. Techniques are also available for detecting and removing cyclical components. Outliers are also removed. What is usually done in time series forecasting for computer performance purposes is to remove the seasonality and the cyclical component to reveal the trend. A curve is then fitted to the trend. The most common curve used is a linear curve but exponential and S-curve fitting is sometimes used as well. After a curve is fitted to the trend data the seasonality and cyclic components are returned to the series so that the forecast can be made. Of course the random component must be taken into account in making the final forecast. Fortunately, we have statistical systems available to handle the rather complex mathematics of all this.

Natural forecasting units are sometimes called business units or key volume indicators because an NFU is usually a business unit. The papers [Browning 1990], [Bowerman 1987], [Reyland 1987], [Lo and Elias 1986], and [Yen 1985] are some of the papers on NFU (business unit) forecasting that have been presented at international CMG conferences. In their paper [Lo and Elias 1986], Lo and Elias list a number of other good NFU forecasting papers.

The basic problem that NFU forecasting solves is that the end users, the people who depend upon computers to get their work done, are not familiar with computer performance units (sometimes called DPUs for data processing units) such as interactions per second, CPU utilization, or I/Os per second, while computer capacity planners are not familiar with the NFUs or the load that NFUs put on a computer system.

Lo and Elias [Lo and Elias 1986] describe a pilot project undertaken to investigate the feasibility of adopting the NFU forecasting technique as part of a capacity planning program. According to Lo and Elias, the major steps needed for applying the NFU forecasting technique are (I have changed the wording slightly from their statement):

1. Identify business elements as possible NFUs.
2. Collect data on the NFUs.
3. Determine the DPUs of interest.
4. Collect the DPU data.
5. Perform the NFU/DPU dependency analysis.
6. Forecast the DPUs from the NFUs.
7. Determine the capacity requirement from the forecasts.
8. Perform an iterative review and revision.

Lo and Elias used the Boole & Babbage Workload Planner software to do the dependency analysis. This software was also used to project the future capacity requirements using standard linear and compound regression techniques. One of their biggest challenges was manually keying in all the data for 266 NFUs. They were able to reduce the number of NFUs to three highly smoothed ones.

Example 7.1

Yen, in his paper [Yen 1985], describes how he predicted future CPU requirements for his IBM mainframe computer installation from input from users. He describes the procedure in the abstract for his paper as follows:

> Projecting CPU requirements is a difficult task for users. However, projecting DASD requirements is usually an easier task. This paper describes a study which demonstrates that there is a positive relationship between CPU power and DASD allocations, and that if a company maintains a consistent utilization of computer processing, it is possible to obtain CPU projections by translating users DASD requirements.

Yen discovered that user departments can accurately predict their magnetic disk requirements (IBM refers to magnetic disks as DASD for "direct access storage device"). They can do this because application developers know the record sizes of files they are designing and the people who will be using the systems can make good predictions of business volumes. Yen used 5 years of historical data describing DASD allocations and CPU consumption in a regression study. He made a scatter diagram in which the y-axis represented CPU hours required for a month, Monday through Friday, 8 am to 4 pm, while the x-axis represented GB of DASD storage installed online on the fifteenth day of that month. Yen found that the regression line $y = 34.58 + 2.59x$ fit the data extraordinarily well. The usual measure of goodness-of-fit is the R-squared value, which was 0.95575. (R-squared is also called the *coefficient of determination.*) In regression analysis studies, R-squared can vary between 0, which means no correlation between x and y values, and ,1 which means perfect correlation between x and y values. A statistician might describe the R-squared value of 0.95575 by saying, "95.575 percent of the total variation in the sample is due to the linear association between the variables x and y." An R-squared value larger than 0.9 means that there is a strong linear relationship between x and y.

Yen no longer has the data he used in his paper but provided me with data from December 1985 through October 1990. From this data I obtained the x and y values plotted in Figure 7.1 together with the regression line obtained from the following *Mathematica* calculations using the standard *Mathematica* package

LinearRegression from the **Statistics** directory of *Mathematica*. The x values are GB of DASD storage online as of the fifteenth of the month, while y is the measured number of CPU hours for the month, normalized into 19 days of 8 hours per day measured in units of IBM System/370 Model 3083 J processors. The Parameter Table in the output from the **Regress** program shows that the regression line is $y = -310.585 + 2.25101\ x$, where x is the number of GB of online DASD storage and y is the corresponding number of CPU hours for the month. We also see that R-squared is 0.918196 and that the estimates of the constants in the regression equation are both considered significant. If you are well versed in statistics you know what the last statement means. If not, I can tell you that it means that the estimates look very good. Further information is provided in the ANOVATable produced by **regress** to bolster the belief that the regression line fits the data very well. However, a glance at Figure 7.1 indicates there are several points in the scatter diagram that appear to be outliers. (An outlier is a data point that doesn't seem to belong to the remainder of the set.) Yen has assured me that the two most prominent points that appear to be outliers really are! The leftmost outlier is the December 1987 value. It is the low point just above the x-axis at $x = 376.6$. Yen says that the installation had just upgraded their DASD so that there was a big jump in installed online DASD storage. In addition, Yen recommends taking out all December points because every December is distorted by extra holidays. The rightmost outlier is the point for December 1989, which is located at (551.25, 627.583). Yen says the three following months are outliers as well, although they don't appear to be so in the figure. Again, the reason these points are outliers is another DASD upgrade and file conversion. We remove all the December points and the other outliers and try again.

```
In[3]:= <<Statistics'LinearRegression'
In[12]:= Regress[data, {1,x}, x]

Out[12]= {ParameterTable ->
    Estimate   SE          TStat       PValue,
 1   -310.585   34.1694     -9.08955    0

 x   2.25101    0.0889939   25.294      0

>     RSquared -> 0.918196, AdjustedRSquared ->
0.91676,

>     EstimatedVariance -> 3684.01,

>     ANOVATable ->
        DoF   SoS              MeanSS          FRatio
```

```
PValue}

Model    1      2.35697 10     2.35697 10     639.785    0

Error    57     209989.        3684.01
                                  6
     Total    58     2.56696 10
```

Figure 7.1. Regression Line for Yen Data

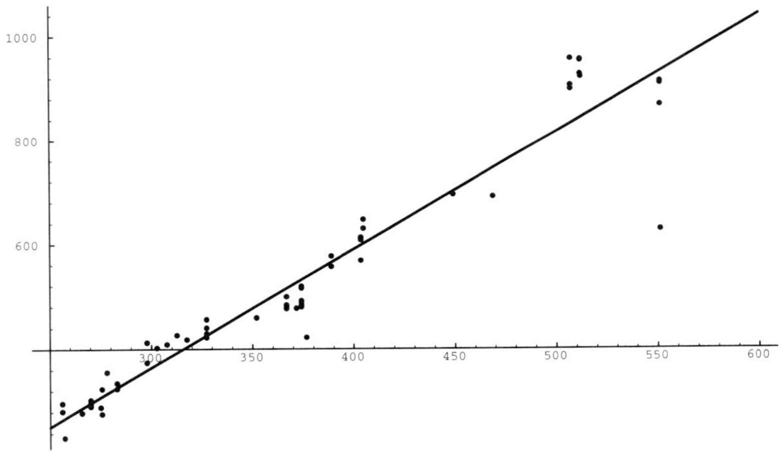

Here we show ParemeterTable from **Regress** for the data with all the outliers, including all December points, deleted:

```
Out[7]= {ParameterTable ->
    Estimate    SE              TStat         PValue,
 1    -385.176    25.6041       -15.0435    0

 x    2.48865     0.0688442    36.149       0

>     RSquared -> 0.963858, AdjustedRSquared ->
0.96312,
 >     EstimatedVariance -> 1478.93,
```

```
>      ANOVATable ->
          DoF    SoS                 MeanSS            FRatio
PValue}
                               6                  6
 Model    1       1.9326 10         1.9326 10       1306.75    0

 Error    49      72467.7           1478.93
                                6
 Total    50     2.00507 10
```

All of the statistical tables got a little scrambled by the capture routine. However, the results are now definitely improved with R-squared equal to 0.963858 and the regression line $y = -385.176 + 2.48865\ x$. The new plot clearly shows the improvement.

Figure 7.2. Regression Line for Corrected Data

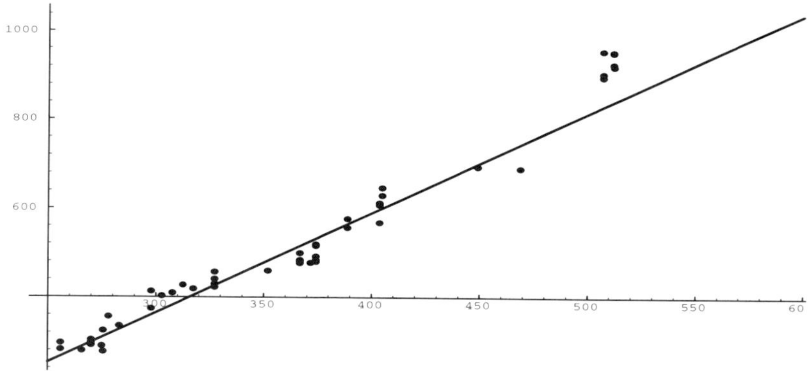

Yen was able to make use of his regression equation plus input from some application development projects to predict when the next computer upgrade was needed. Let us examine how that might be done with the data in Figure 7.2. The rightmost data point is (512.15, 921.019). Since there are 152 hours in a time period consisting of 19 days with 8 hours per day, the number of equivalent IBM 3083 Model J CPUs for this point is 6.06. We assume that Blue Cross has the equivalent of at least 7 IBM 3083 Model J computers at this time. If it is exactly 7, we would like to know when at least 8 will be needed. We can use the regression line to estimate this as shown in the following *Mathematica* calculation. We see that at least eight equivalent CPUs will be needed when the online storage reaches 643.391 GB. We can predict when that will happen and thus when an upgrade will be needed, at least to within a few months.

```
In[58]:= f[x] = -385.176 + 2.48865 x

Out[58]= -385.176 + 2.48865 x

In[59]:= Solve[f[x]/152 == 8.0, x]

Out[59]= {{x -> 643.391}}
```

While the technique used by Yen to predict when the next upgrade should occur within a few months, forecasting of total CPU hours needed per month alone does not provide much information on the performance of the system as it approaches the point where more computing capacity is needed. More detailed information is needed to determine when the performance deteriorates so that the users feel that such performance measures as average response time are unacceptable. Yen and his colleagues of course tracked performance information to avoid this problem. In fact, Yen used the modeling package Best/1 MVS to make frequent performance predictions. The forecasting process allowed Yen to predict far in advance when an upgrade would likely be needed so that the necessary procurement procedures could be carried out in a timely fashion.

Exercise 7.1

Apply linear regression to the file **data1** that is on the diskette in the back of the book. Hint: Don't forget to read in the package **LinearRegression** from **Statistics**. How you read it in depends upon what version of *Mathematica* you have.

Example 7.2

This example is taken from the HP RXForecast User's Manual For HP-UX Systems. One of the useful features of HP RXForecast is the capability of associating business units with computer performance metrics to see if there is a correlation. When there is a strong correlation, HP RXForecast will forecast computer performance metrics from business unit forecasts. For this example the **scopeux** collector was run continuously from January 3, 1990, until March 19, 1990, to generate the TAHOE.PRF performance log file. Then HP RXForecast was used to correlate the global CPU utilization to the business units provided in the business unit file TAHOEWK.BUS. The flat ASCII file called

TAHOEWK.BUS shown in Table 7.1 represents the amount of work completed each week in business units.

Table 7.1. Business Unit File

Month	Week	Year	Units
1	1	1990	2800
1	2	1990	5510
1	3	1990	4300
1	4	1990	5000
2	1	1990	5920
2	2	1990	4800
2	3	1990	3000
2	4	1990	5700
3	1	1990	4800
3	2	1990	5200
3	3	1990	7800
3	4	1990	6500
4	1	1990	6700
4	2	1990	7000
4	3	1990	6200
4	4	1990	7400
5	1	1990	7700
5	2	1990	6900
5	3	1990	8100
5	4	1990	8300
6	1	1990	8600

Table 7.1. Business Unit File (Continued)

6	2	1990	8100
6	3	1990	9000
6	4	1990	9300

Figure 7.3. Business Unit Forecasting Example

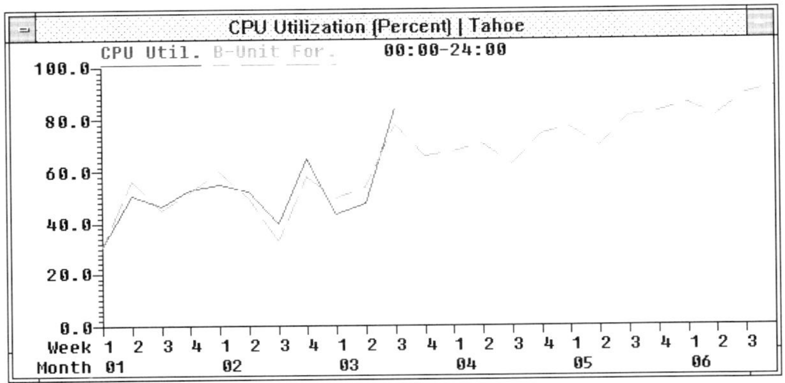

The graph shown in Figure 7.3 was produced by HP RXForecast. The first part of the graph (up to week 3 of the third month) compares the actual global CPU utilization and the global CPU utilization predicted by regression of CPU utilization on business units. The two curves are very close. The single curve starting in the third week of the third month is the RXForecast forecast of CPU utilization from the predicted business units. The regression for the first part of the curve is very good with an *R*-squared value of 0.86 and a standard error of only 5.49. Note that, for the business unit forecasting technique to work, the prediction of the growth of business units must be provided to HP RXForecast.

7.3 Solutions

Solution to Exercise 7.1

We used *Mathematica* as shown here except we that do not indicate how the data1 file was read by a simple <<data1 because it dumps all the numbers on the screen. We also display only the final graphic. The fit looks pretty good in Figure 7.4 although the *R*-squared value of 0.883297 is slightly lower than we'd like.

```
In[3]:= <<Statistics'LinearRegression'
In[6]:= gp = ListPlot[data]

Out[6]= -Graphics-

In[7]:= Regress[data, {1, x}, x]

Out[7]= {ParameterTable -> Estimate SE TStat PValue ,
 1 -252.609 48.1013 -5.2516 0.0000287096

 x 2.08306 0.161428 12.904 0

> RSquared -> 0.883297, AdjustedRSquared -> 0.877992,

> EstimatedVariance -> 579.745,

> ANOVATable -> DoF SoS MeanSS FRatio PValue}
 Model 1 96534.6 96534.6 166.512 0

 Error 22 12754.4 579.745
Total 23 109289.

In[8]:= g = Fit[data, {1, x}, x]

Out[8]= -252.609 + 2.08306 x

In[12]:= gg = Plot[g, {x, 240, 390}]

Out[12]= -Graphics-

In[13]:= Show[gg, gp]
```

Figure 7.4. Output from Exercise 7.1

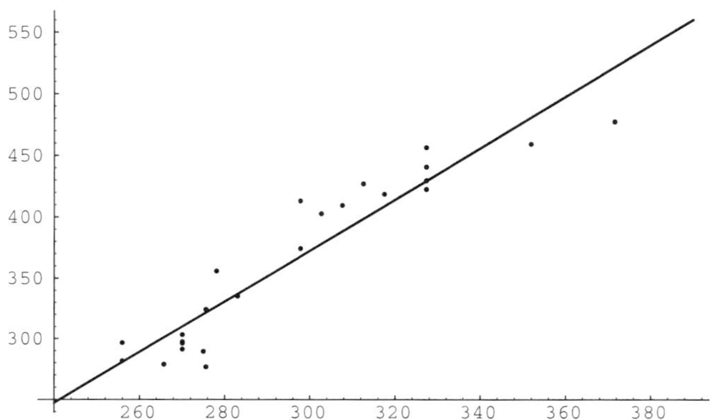

7.4 References

1. Tim Browning, "Forecasting computer resources using business elements: a pilot study," *CMG '90 Conference Proceedings*, Computer Measurement Group, 1990, 421—427.

2. James R. Bowerman, "An introduction to business element forecasting," *CMG '87 Conference Proceedings,* Computer Measurement Group, 1987, 703—709.

3. C. Chatfield, *The Analysis of Time Series: An Introduction,* Third Edition, Chapman and Hall, London, 1984.

4. T. L. Lo and J. P. Elias, "Workload forecasting using NFU: a capacity planner's perspective," *CMG '86 Conference Proceedings,* Computer Measurement Group, 1986, 115—120.

5. George W. (Bill) Miller, "Workload characterization and forecasting for a large commercial environment," *CMG '87 Conference Proceedings,* Computer Measurement Group, 1987, 655—665.

6. John M. Reyland, "The use of natural forecasting units," *CMG '87 Conference Proceedings,* Computer Measurement Group, 1987, 710—13.

7. Kaisson Yen, "Projecting SPU capacity requirements: a simple approach," *CMG '85 Conference Proceedings*, Computer Measurement Group, 1985, 386—391.

Chapter 8 Afterword

The reasonable man adapts himself to the world; the unreasonable one persists in trying to adapt the world to himself. Therefore all progress depends on the unreasonable man.

George Bernard Shaw

8.1 Introduction

I hope the reader fits Shaw's definition of "unreasonable" and wants to change things for the better. The purpose of this chapter is to review the first seven chapters of this book and to suggest what you might do to continue your education in computer performance analysis.

8.2 Review of Chapters 1—7

8.2.1 Chapter 1: Introduction

In Chapter 1 we supply definitions and descriptions of the concepts and techniques used in computer performance analysis. We also provide an overview of the book and a discussion of the management techniques required for managing the performance of a computer system or systems. These management techniques include the use of *service level agreements* (SLAs), *chargeback systems*, and the use of *capacity planning*. Capacity planning has both management and technical components. The *service level agreement*, a contract between the provider of the service (we will call this entity IS for Information Systems here) and the end users, is a key management technique. It requires the two groups to engage in a dialogue so that mutually acceptable performance requirements can be set.

Installations sometimes use chargeback in conjunction with SLAs so that user organizations are more aware of the fact that improved performance often requires increased costs. (A familiar adage here is, "There ain't no free lunch.")

To carry out the requirements of an SLA, IS must use other techniques described in Chapter 1. The main technique that must be mastered is capacity planning. The purpose of capacity planning is to provide an acceptable level of computer service to the organization while responding to workload demands generated by business requirements. Thus IS must forecast (predict) future workload, predict when upgrades are required, and predict the performance of possible future configurations. (Capacity planning is needed even when there are no service level agreements.) The discipline needed for evaluating the performance of proposed configurations is called *performance prediction*; modeling is the main tool used in this discipline.

The modeling techniques available for performance prediction include rules of thumb, back-of-the-envelope calculations, statistical forecasting, analytical queueing theory, simulation, and benchmarking. We provide an overview of each of these techniques in Chapter 1 with examples of how they might be used. We also provide trade-offs to help you decide which technique (or techniques) is the best for your installation. The more complex techniques are discussed in more depth in later chapters.

Software performance engineering (SPE) is an important concept that has recently appeared. It is a method to help software developers ensure that application software will meet performance goals at the end of the development cycle.

Another important topic discussed in Chapter 1 is performance management tools. We discuss a number of tools and provide examples of the output from representative examples of these tools. One of the leading edge performance management tools is the expert system for computer performance analysis. This tool is particularly important at computer installations with no experienced performance experts or for very complex operating systems such as the IBM MVS/XA or MVS/ESA operating systems for IBM or compatible mainframes; MVS is so complex that even the experts have trouble keeping up with all the latest changes and recommendations.

We close Chapter 1 by discussing organizations and journals that are important for computer performance analysts.

8.2.2 Chapter 2: Components of Computer Performance

In chapter 2 we discuss the components of computer performance. We begin this discussion by defining exactly what is meant by the statement, "machine A is n% faster than machine B in performing task X." It is defined by the formula

$$\frac{\text{Execution Time}_B}{\text{Execution Time}_A} = 1 + \frac{n}{100},$$

where the numerator in the fraction is the time it takes machine B to execute task X and the denominator is the time it takes machine A to do so. Solving for *n* yields

$$n = \frac{\text{Execution Time}_B - \text{Execution Time}_A}{\text{Execution Time}_A} \times 100.$$

We provide the *Mathematica* program **perform** in the package **first.m** to make this calculation.

Another important formula is known as *Amdahl's law* and tells us the speedup that can be achieved by improving the performance of part of a computer system such as a CPU or an I/O device. The formula for Amdahl's law is given by

$$\frac{\text{Execution Time}_{\text{old}}}{\text{Execution Time}_{\text{new}}} = \frac{1}{1 - \text{Fraction}_{\text{enhanced}} + \dfrac{\text{Fraction}_{\text{enhanced}}}{\text{Speedup}_{\text{enhanced}}}} = \text{Speedup}_{\text{overall}}.$$

This formula defines speedup and describes how we calculate it using Amdahl's law, the middle formula. Thus the speedup is two if the new execution time is one half the old execution time. Amdahl's law shows that, if one quarter of the execution time of a job is spent doing I/O, which is then enhanced to run twice as fast, the resulting overall speedup is 8/7 or 1.143.

The *Mathematica* program **speedup** in the package **first.m** can be used to make this calculation.

Processors (CPUs)

One of the most important components of any computer system is the central processing unit (CPU) (CPUs on multiprocessor systems). The processing power of a CPU is primarily determined by the *clock cycle* or smallest unit of time in which the CPU can execute a single instruction. (According to [Kahaner and Wattenberg 1992] the Hitachi S-3800 has the shortest clock cycle of any commercial computer in the world; it is two billionths of a second!) Some *superscalar* RISC (reduced instruction set computer) systems can execute more than one instruction per cycle by *pipelining*. Pipelining is a method of improving the throughput of a CPU by overlapping the execution of multiple instructions. It is described in detail in [Hennessy and Patterson 1990] and conceptually in [Denning 1993]. It is customary to provide basic CPU speed in units of millions of clock cycles per second or MHz. As this is being written (June 1993) the fastest microprocessor available on an IBM PC or compatible is the 66 MHz Intel Pentium. An unfortunate name that is sometimes attached to CPU speed is the MIPS or millions of instructions executed per second. MIPS is a poor measure of CPU performance because the number of instructions per second executed by any computer depends very much on exactly what kind of work the computer is doing; this is true because different instructions require different execution times. Thus a floating point multiplication generally requires more time to execute than a fixed point addition. The obvious solution is to measure MIPS on all machines by having having them execute exactly the same program. Alas, this approach does not work either because machines with different architectures and thus different instruction sets execute different numbers of instructions in executing the same program. Another unsuccessful approach is to declare one machine a standard (the VAX-11/780 is the most common example) and compare the time it takes to perform a certain task against the time it takes to perform the same task on the standard machine thus generating Relative MIPS. At one time the VAX-11/780 was thought to be a 1 MIPS machine. It is now known to be approximately a 0.5 MIPS machine. By this we mean that for most programs run on the VAX 11/780 it executes approximately 500,000 instructions per second. When a computer manufacturer says one of the computers it sells is a 50 MIPS machine, it usually means 50 Relative VAX MIPS and is commonly computed by running the Dhrystone 1.1 benchmark to obtain a Dhrystones per second rating; this number is then divided by 1,757 to obtain the number of Relative VAX MIPS. The Dhrystone benchmark was developed by Weicker in 1984 to measure the performance of system programming types of operating systems, compilers, editors, etc. The result of running the Dhrystone benchmark is reported in Dhrystones per second. Weicker in his paper [Weicker 1990] describes his original benchmark as well as Versions 1.1 and 2.0. According to a well-known PC performance measurement tool, my 33 MHz 80486DX IBM PC compatible has a relative VAX MIPS rating of 14.652.

The total time required for a CPU to execute a sequence of instructions is given by the formula

$$\text{CPU time} = \text{Instruction count} \times \text{CPI} \times \text{Clock cycle time.}$$

where the first variable on the right is the total number of instructions executed, CPI is the average number of clock cycles needed to execute a CPU instruction, and the last variable is the clock cycle time. The *Mathematica* program **cpu** in the package **first.m** utilizes the three inputs: (1) number of instructions executed, (2) CPU clock rate in MHz, and (3) time in seconds taken to execute the given instructions. It produces the CPI and the MIPS for the calculation. For example, as we show in Chapter 2, if a 50 MHz CPU executes 750 million instructions in

50 seconds, the CPI is 3 1/3 clock cycles per instruction, and the MIPS rating is 15 for the code executed. Both of these numbers would probably be different if a different code sequence was executed.

Multiprocessors

Many computer systems have more than one processor (CPU) and thus are known as *multiprocessor systems*. There are two basic organizations for such systems: loosely coupled and tightly coupled.

Tightly coupled multiprocesors, also called shared memory multiprocessors, are distinguished by the fact that all the processors share the same memory. There is only one operating system, which synchronizes the operation of the processors as they make memory and data base requests. Most such systems allow a certain degree of parallelism; that is, for some applications they allow more than one processor to be active simultaneously doing work for the same application. Tightly coupled multiprocessor computer systems can be modeled using queueing theory and information from a software monitor. This is a more difficult task than modeling uniprocessor systems because of the interference between processors. Modeling is achieved using a load dependent queueing model together with some special measurement techniques.

Loosely coupled multiprocessor systems, also known as *distributed memory systems*, are sometimes called *massively parallel computers* or *multicomputers*. Each processor has its own memory and sometimes a local operating system as well. There are several different organizations for loosely coupled systems but the problem all of them have in achieving high speeds is indicated by Amdahl's law, which says that the degree of speedup due to the parallel operation is given by

$$\text{Speedup} = \frac{1}{(1 - \text{Fraction}_{\text{parallel}}) + \dfrac{\text{Fraction}_{\text{parallel}}}{n}},$$

where n is the total number of processors. The problem is achieving a high degree of parallelism. For example, if the system has 100 processors with all of them running in parallel one half of the time, the speedup is only 1.9802. To obtain a speedup of 50 requires that the fraction of the time that all processors are operating in parallel is 98/99=0.98989899.

We discuss a number of the leading multiprocessor computer systems in Chapter 2. We also recommend the September 1992 issue of *IEEE Spectrum*. It is a special issue devoted to supercomputers and it covers all aspects of the newest computer architectures as well as the problems of developing software to take advantage of the processing power.

The memory hierarchy is another important component of computer performance

Figure 8.1. The Memory Hierarchy

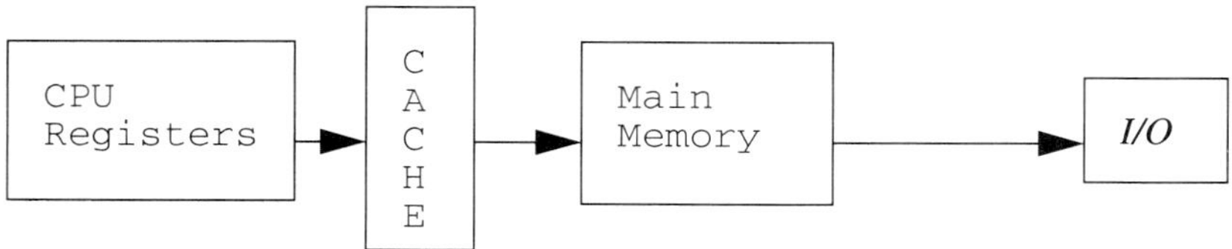

Figure 8.1 shows the typical memory hierarchy on a computer system; it is valid for most computers ranging from personal computers and workstations to supercomputers. The fastest memory, and the smallest in the system, is provided by the CPU registers. As we proceed from left to right in the hierarchy memories become larger, the access times increase, and the cost per byte decreases. The goal of a well-designed memory hierarchy is a system in which the average memory access times are only slightly slower than that of the fastest element, the CPU cache (the CPU registers are faster than the CPU cache but cannot be used for general storage), with an average cost per bit that is only slightly higher than that of the lowest cost element.

A CPU cache is a small, fast memory that holds the most recently accessed data and instructions from main memory. Some computer architectures, such as the Hewlett-Packard Precision Architecture, call for separate caches for data and instructions. When the item sought is not found in the cache, a *cache miss* occurs, and the item must be retrieved from main memory. This is a much slower access, and the processor may become idle while waiting for the data element to be delivered. Fortunately, because of the strong locality of reference exhibited by a program's instruction and data reference sequences, 95 to more than 98 percent of all requests are satisfied by the cache on a typical system. Caches work because of the *principle of locality*. This concept is explained in great detail in the excellent book [Hennessy and Patterson 1990]. A cache operates as a system that moves recently accessed items and the items near them to a storage medium that is faster than main memory.

Just as all objects referenced by the CPU need not be in the CPU cache or caches, not all objects referenced in a program need be in main memory. Most computers (even personal computers) have virtual memory so that some lines of a program may be stored on a disk. The most common way that virtual memory is handled is to divide the address space into fixed-size blocks called pages. At any give time a page can be stored either in main memory or on a disk. When the CPU references an item within a page that is not in the CPU cache or in main memory, a *page fault* occurs, and the page is moved from disk to main memory. Thus the CPU cache and main memory have the same relationship as main memory and disk memory. Disk storage devices, such as the IBM 3380 and 3390, have cache storage in the disk control unit so that a large percentage of the time a page or block of data can be read from the cache obviating the need to perform a disk read. Special algorithms and hardware for writing to the cache have also been developed. According to Cohen, King, and Brady [Cohen, King, and Brady 1989] disk cache controllers can give up to an order of magnitude better I/O service time than an equivalent configuration of uncached disk storage.

Because caches consist of small, speedy memory elements they are very fast and can significantly improve the performance of computer systems. In Chapter 2 we give some examples of how CPU caches can improve performance.

Input and output is a very important component of the performance of computer systems although this fact is frequently overlooked. The most important I/O device for most computers is the magnetic disk drive, which we discuss is some detail in Chapter 2.

The hottest new innovation in disk storage technology is the *disk array*, more commonly denoted by the acronym *RAID* (Redundant Array of Inexpensive Disks). The seminal paper for this technology is the paper [Patterson, Gibson, and Katz 1988]. It introduced RAID terminology and established a research agenda for a group of researchers at UC Berkeley for several years. The abstract of their paper, which provides a concise statement about the technology follows.

> Increasing performance of CPU and memories will be squandered if not matched by a similar performance increase in I/O. While the capacity of Single Large Expensive Disks (SLED) has grown rapidly, the performance improvement of SLED has been modest. Redundant Arrays of Inexpensive Disks (RAID), based on the magnetic disk technology developed for personal computers, offers an attractive alternative to SLED, promising improvements of an order of magnitude in performance, reliability, power consumption, and scalability. This paper introduces five levels of RAID, giving their relative cost/

performance, and compares RAID to an IBM 3380 and a Fujitsu Super Eagle.

RAID is a new technology. In Chapter 2 we discuss some of the considerations of using this form of I/O.

In the final section of Chapter 2 we discuss the interplay between CPUs, I/O, and memory as it affects performance.

8.2.3 Chapter 3: Basic Calculations

In Chapter 3 we introduce the basic queueing network models that are used for most modeling studies of computer performance. For all performance calculations we assume some sort of model of the system under study. A model is an abstraction of a system that is easier to manipulate and experiment with than the real system—especially if the system under study does not yet exist. It could be a simple back-of-the-envelope model. However, for more formal modeling studies, computer systems are usually modeled by symbolic mathematical models. We usually use a queueing network model when thinking about a computer system. The most difficult part of effective modeling is determining what features of the system must be included and which can safely be left out. Fortunately, using a queueing network model of a computer system helps us solve this key modeling problem. The reason for this is that queueing network models tend to mirror computer systems in a natural way. Such models can then be solved using analytic techniques or by simulation. In this chapter we will show that quite a lot can be calculated using simple back-of-the envelope techniques. These are made possible by some queueing network laws including Little's law, the utilization law, the response time law, and the forced flow law. In Chapter 3 we illustrate these laws with examples and provide some simple exercise to enable you to test your understanding.

When we think of a computer system a model similar to Figure 8.2 comes to mind. We think of people at terminals or workstations making requests for computer service such as entering a customer purchase order, finding the status of a customer's account, etc. The request goes to the computer system where there may be a queue for memory before the request is processed. As soon as the request enters main memory and the CPU is available it does some processing of the request until an I/O request is required; this may be due to a page fault (the CPU references an instruction that is not in main memory) or to a request for data. When the I/O request has been processed the CPU continues processing of the original request between I/O requests until the processing is complete and a response is sent back to the user's terminal. This model is a queueing network model which can be solved using either analytic queueing theory or simulation.

Figure 8.2. Closed Computer System

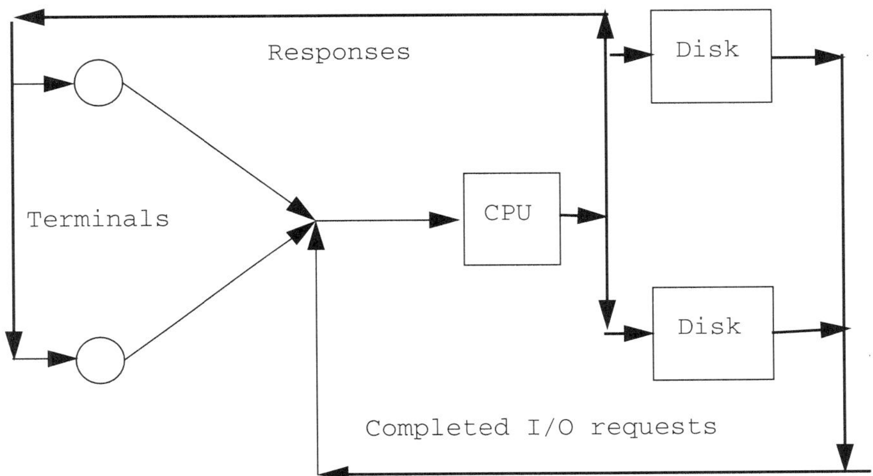

The queueing network model view of a computer system is that of a collection of interconnected service centers and a set of customers who circulate through the service centers to obtain the service they require as we indicate in Figure 8.1. Thus to specify the model we must define the customer service requirements at each of the service centers, as well as the number of customers and/or their arrival rates. This latter description is called *workload intensity.* Thus workload intensity is a measure of the rate at which work arrives for processing.

In Chapter 3 we discuss single workload class models in which all users of the computer system are assumed to be performing the same application as well as the more common system in which different types of workloads are executed simultaneously.

Workload types are defined in terms of how the users interact with the computer system. Some users employ terminals or workstations to communicate with their computer system in an interactive way. The corresponding workload is called a *terminal workload.* Other users run batch jobs, that is, jobs that take a relatively long time to execute. In many cases this type of workload requires special setup procedures such as the mounting of tapes or removable disks. For historical reasons such workloads are called *batch workloads*. The third kind of workload is called a *transaction workload* and does not correlate quite so closely with the way an actual user utilizes a computer system. Large data base systems such as airline reservation systems have transaction workloads, which correspond roughly to computer systems with a very large number of active terminals.

There are two types of parameters for each workload type: parameters that specify the workload intensity and parameters that specify the service requirement of the workload at each of the computer service centers.

We describe the workload intensity for each of the three workload types as follows:

1. The intensity of a terminal workload is specified by two parameters: N, the average number of active terminals (users), and Z, the average think time. The think time is the time between the response to a request and the start of the next request.

2. The intensity of a batch workload is specified by the parameter N, the average number of active customers (transactions or jobs). Batch workloads have a fixed population. Batch jobs that complete service are thought of as leaving the system to be replaced instantly by a statistically identical waiting job. Thus a batch workload could have an intensity of $N = 6.2$ jobs so that, on the average, 6.2 of these jobs are running on the computer system.

3. A transaction workload intensity is given by λ, the average arrival rate of customers (requests). Thus it has the dimensions of customers divided by time, such as 1,000 inquiries per hour or 50 transactions per second. The population of a transaction workload that is being processed by the computer system varies over time. Customers leave the system upon completing service.

A queueing model with a transaction workload is an open model since there is an infinite stream of arriving and departing customers. When we think of a transaction workload we think of an open system as shown in Figure 8.3 in which requests arrive for processing, circulate about the computer system until the processing is complete, and then leave the system. Conversely, models with batch or terminal workloads are called *closed* models since the customers can be thought of as never leaving the system but as merely recirculating through the system as shown in Figure 8.2. We treat batch and terminal workloads the same from a modeling point of view; batch workloads are terminal workloads with think time zero. As we will see later, using transaction workloads to model some computer systems can lead to egregious errors. We recommend *fixed throughput workloads* instead. They are discussed in Chapter 4.

Figure 8.3. Open Computer Model

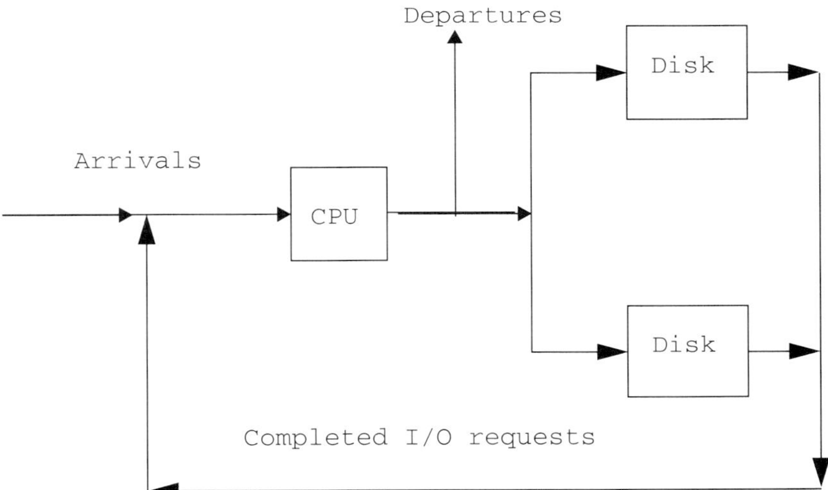

The only difference in nomenclature for models with multiple workload classes rather than a single workload class is that each workload parameter must be indexed with the workload number. Thus a terminal class workload has the parameters N_c and Z_c as well as the average service time per visit $S_{c,k}$ and the average number of visits required $V_{c,k}$ for each service center k.

A queueing network is a collection of service centers connected together so that the output of any service center can be the input to another. That is, when a customer completes service at one service center the customer may proceed to another service center to receive another type of service. Here we are following the usual queueing theory terminology of using the word "customer" to refer to a service request. For modeling an open computer system we have in mind a queueing network similar to that in Figure 8.3.

In Figure 8.3 the customers (requests for service) arrive at the computer center where they begin service with a CPU burst. Then the customer goes to one of the I/O devices (disks) to receive some I/O service (perhaps a request for a customer record). Following the I/O service the customer returns to the CPU queue for more CPU service. Eventually the customer will receive the final CPU service and leave the computer system.

We assume that the queueing network representation of a computer system has C customer classes and K service centers. We use the symbol $S_{c,k}$ for the average service time for a class c customer at service center k, that is, for the average time required for a server in service center k to provide the required service to one class c customer. It is the reciprocal of $\mu_{c,k}$, which is a Greek symbol used to represent the average service rate or the average number of class c customers serviced per unit of time at service center k when the service center is busy.

The average response time, R, and average throughput, X, are the most common system performance metrics for terminal and batch workloads. These same performance metrics are used for queueing networks, both as measurements of system wide performance and measurements of service center performance. In addition we are interested in the average utilization, U, of each service facility. For any server the average utilization of the device over a time period is the fraction of the time that the server is busy. Thus, if over a 10 minute period the CPU is busy 5 minutes, then we have $U = 0.5$ for that period. Sometimes the utilization is given in percentage terms so this utilization would be stated as 50% utilization. In Chapter 3 we discuss the queueing network performance measurements separately for single workload class models and multiple workload class models. For single workload class models, the primary system performance parameters are the average response time, R, the average throughput, X, and the average number of customers in the system, L. In addition, for each service center we are interested in the average utilization, the average time a customer spends at the center, the average center throughput, and the average number of customers at the center.

For multiple workload class models there also are system performance measures and center performance measures. Thus we may be interested in the average response time for users who are performing order entry as well as for those who are making customer inquiries. In addition we may want to know the breakdown of response time into the CPU portion and the I/O portion so that we can determine where upgrading is most urgently needed.

Similarly, we have service center measures of two types: aggregate or total measures and per class measures. Thus we may want to know the total CPU utilization as well as the breakdown of this utilization between the different workloads.

Queueing Network Laws

One of the most important topics discussed and illustrated with examples in Chapter 3 is queueing network laws. The single most profound and useful law of computer performance evaluation (and queueing theory) is called **Little's law** after John D. C. Little who gave the first formal proof in his 1961 paper [Little 1961]. Before Little's proof the result had the status of a folk theorem, that is, almost everyone believed the result was true but no one knew how to prove it. The use of Little's law is the most important and useful principle of queueing theory and his paper is the single most quoted paper in the queueing theory literature.

Little's law applies to any system with the following properties:

1. Customers enter and leave the system.
2. The system is in a steady-state condition in the sense that $\lambda_{in} = \lambda_{out}$ where λ_{in} is the average rate that customers enter the system and λ_{out} is the average rate that customer leave the system.

Then, if $X = \lambda_{in} = \lambda_{out}$, L is the average number of customers in the system, and R is the average amount of time each customer spends in the system, we have the relation $L = X \times R$.

Thus Little's law provides a relationship between the three variables L, X, and R. The relationship can be written in two other equivalent forms: $X = L/R$, and $R = L/X$.

One of the corollaries of Little's law is the **utilization law.**

It relates the throughput X, the average service time S, and the utilization U of a service center by the formula $U = X \times S$.

Consider Figure 8.2. Assume this is a closed single workload class model of an interactive system with N active terminals, and a central computer system with one CPU and some I/O devices. Little's law can be applied to the whole system to discover the relation between the throughput X, the average think time Z, the response time R, and the number of terminals N. The result is the **response time law**

$$R = \frac{N}{X} - Z.$$

The response time law can to generalized to the multiclass case to yield

$$R_c = \frac{N_c}{X_c} - Z_c.$$

In Section 3.3.3 we provide several examples of the use of the response time law.

For a single workload class computer system the **forced flow law** says that the throughput of service center k, X_k, is given by $X_k = V_k \times X$ where X is the computer system throughput. This means that a computer system is holistic in the sense that the overall throughput of the system determines the throughput through each service center and vice versa.

We repeat Example 3.3 below (as Example 8.1) because it illustrates several of the laws under discussion

Example 8.1

Suppose Arnold's Armchairs has an interactive computer system (single workload) with the characteristics shown in Table 8.1.

:

Table 8.1. Data for Example 8.1

Parameter	Description
$N = 10$	There are 10 active terminals
$Z = 18$	Average think time is 18 seconds
$V_{disk} = 20$	Average number of visits to this disk is 20 per interaction
$U_{disk} = 0.25$	Average disk utilization is 25 percent
$S_{disk} = 0.25$ seconds	Average disk service time per visit is 0.25 seconds

We make the following calculations:

Since, by the utilization law, $U_{disk} = X_{disk} \times S_{disk}$, we calculate

$$X_{disk} = \frac{U_{disk}}{S_{disk}} = \frac{0.25}{0.025} = 10$$

requests per second.

We can rewrite the forced flow law as $X = X_k/V_k$. Hence, the average system throughput is given by $X = 10/20 = 0.5$ interactions per second. By the response time law we calculate the average response time as $R = 10/0.5 - 18 = 2.0$ seconds.

One of the key performance concepts used in studying a computer system is the *bottleneck device* or *server*, usually referred to as the *bottleneck*. The name derives from the neck of a bottle which restricts the flow of liquid. As the workload on a computer system increases some resource of the system eventually becomes overloaded and slows down the flow of work through the computer. The resource could be a CPU, an I/O device, memory, or a lock on a data base. When this happens the combination of the saturated resource (server) and a randomly changing demand for that server causes response times and queue lengths to grow dramatically. By *saturated server* we mean a server with a utilization of 1.0 or 100%. A *system is saturated* when at least one of its servers or resources is saturated. The *bottleneck* of a system is the first server to saturate as the load on the system increases. Clearly, this is the server with the largest total service demand.

It is important to note that the bottleneck is workload dependent. That is, different workloads have different bottlenecks for the same computer system. It is part of the folklore that scientific computing jobs are *CPU bound*, while business oriented jobs are *I/O bound*. That is, for scientific workloads such as CAD (computer aided design), FORTRAN compilations, etc., the CPU is usually the bottleneck. Business oriented workloads, such as data base management systems, electronic mail, payroll computations, etc., tend to have *I/O* bottlenecks. Of course, one can always find a particular scientific workload that is not CPU bound and a particular business system that is not *I/O* bound, but it is true that different workloads on the same computer system can have dramatically different bottlenecks. Since the workload on many computer systems changes during different periods of the day, so do the bottlenecks. Usually, we are most interested in the bottleneck during the peak (busiest) period of the day.

Chapter 3 is rounded out by a discussion of bounds for queueing systems, a discussion of the modeling study paradigm, and a discussion of why queueing theory models are important for performance calculations.

The bounds are useful for back-of-the-envelope calculations, a review of the modeling study paradigm is important because many modeling studies are undertaken without a clear statement of objectives, and there is a bias against queueing models in some quarters because of a fear of mathematics.

8.2.4 Chapter 4: Analytic Solution Methods

In Chapter 4 we discuss the mean value analysis (MVA) approach to the analytic solution of queueing network models. MVA is a solution technique developed by Reiser and Lavenberg in [Reiser 1979, Reiser and Lavenberg 1980]. In Chapter 6 we discuss solutions of queueing network models through simulation.

Although analytic queueing theory is very powerful there are queueing networks that cannot be solved exactly using the theory. In their paper [Baskett, Chandy, Muntz, and Palacios 1975], a widely quoted paper in analytic queueing theory, Baskett *et al.* generalized the types of networks that can be solved analytically. Multiple customer classes, each with different service requirements, as well as service time distributions other than exponential are allowed. Open, closed, and mixed networks of queues are also allowed. They allow four types of service centers, each with a different queueing discipline. Before this seminal paper was published most queueing theory was restricted to Jackson networks that allowed only one customer class (a single workload class) and required all service times to be exponential. The exponential distribution is a popular one in applied probability because of its nice mathematical properties and because many real world probability distributions are approximately exponential. The networks described by Baskett *et al.* are now known as *BCMP networks*. For these networks efficient solution algorithms are known; many of them are presented in Chapter 4 together with *Mathematica* programs for their solution.

Single Class Workload Models

We begin by showing how to solve single workload class models because these models are very easy to solve and the solution techniques are fairly straightforward, especially for open models. The open, single class model is an approximate model, since there is no actual open, single class computer system. The equations for this model are displayed in Table 4.1 and implemented by the *Mathematica* program **sopen** in the package **work.m.** We provide an example and several exercises using this model. The closed single class model is more complex; we provide the description of the MVA algorithm for this model from Chapter 4 below.

We visualize a closed single class model in Figure 8.4. The N terminals are treated as delay centers. We assume that the CPU is either an exponential server with the FCFS queue discipline or a processor sharing (PS) server. By FCFS queueing discipline we mean that customers are served in the order in which they arrive. Processor sharing is a generalization of round-robin in which each customer shares the server equally. Thus, for a processor sharing server, if there are five customers at the server each of them receives one fifth of the power of the server.

The I/O devices are all treated as having the FCFS queue discipline. We assume that the CPU and I/O devices are numbered from 1 to K with the CPU counted as device 1. The MVA algorithm for the performance calculations follows.

Single Class Closed MVA Algorithm. Consider the closed computer system of Figure 8.4. Suppose the mean think time is Z for each of the N active terminals. The CPU has either the FCFS or the processor sharing queue discipline with service demand D_1 given. We are also given the service demands of each I/O device numbered from 2 to K. We calculate the performance measures as follows:

Step 1 [Initialize] Set $L_k[0] = 0$ for $k = 1, 2, \ldots, K$.
Step 2 [Iterate] For $n = 1, 2, \ldots, N$ calculate

$$R_k[\mathrm{n}] = D_k(1 + L_k[\mathrm{n}\text{-}1]),\ k = 1, 2, \ldots, K,$$

$$R[n] = \sum_{k=1}^{K} R_k[\mathrm{n}],$$

$$X[n] = \frac{n}{R[\mathrm{n}] + Z},$$

$$L_k[\mathrm{n}] = X[\mathrm{n}]\mathrm{R}_k[\mathrm{n}],\ \ \mathrm{k}{=}1{,}2{,}\ldots,K.$$

Step 3 [Compute Performance Measures] Set the system throughput to

$$X = X[N].$$

Set response time (turnaround time) to

$$R = R[N].$$

Set the average number of customers (jobs) in the main computer system to $L = X\mathrm{R}$.

Set server utilizations to $U_k = X\,\mathrm{D}_k$, k=1,2, ..., K.

We calculated $L_k[\mathrm{N}]$ and $R_k[\mathrm{N}]$ for each server in the last iteration of Step 2.

Figure 8.4. Closed MVA Model

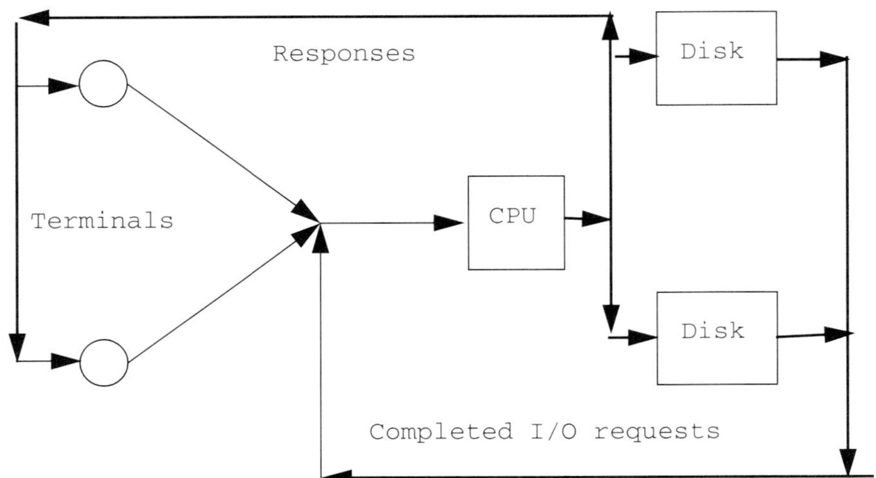

The algorithm is actually quite straightforward and intuitive except for the first equation of Step 2 which depends upon the *arrival theorem*, stated by Reiser in [Reiser 1981] as follows:

> In a closed queueing network the (stationary) state probabilities at customer arrival epochs are identical to those of the same network in long-term equilibrium with one customer removed.

Like all MVA algorithms, this algorithm depends upon Little's law (discussed in Chapter 3), and the arrival theorem. The key equation is the first equation of Step 2, $R_k[\text{n}] = D_k(1 + L_k[\text{n-1}])$, which is executed for each service center. By the arrival theorem, when a customer arrives at service station k the customer finds $L_k[\text{n-1}]$ customers already there. Thus the total number of customers requiring service, including the new arrival, is $1 + L_k[\text{n-1}]$. Hence the total time the new customer spends at the center is given by the first equation in Step 2 if we assume we needn't account for the service time that a customer in service has already received. The fact that we need not do this is one of the theorems of MVA! The arrival theorem provides us with a bootstrap technique needed to solve the equation $R_k[\text{n}] = D_k(1 + L_k[\text{n-1}])$ for $n = N$. When n is 1 $L_k[\text{n-1}] = L_k[0] = 0$

so that $R_k[1] = D_k$, which seems very reasonable; when there is only one customer in the system there cannot be a queue for any device so the response time at each device is merely the service demand. The next equation is the assertion that the total response time is the sum of the times spent at the devices. The last two equations are examples of the application of Little's law. The final equation provides the input needed for the first equation of Step 2 for the next iteration and the bootstrap is complete. Step 3 completes the algorithm by observing the performance measures that have been calculated and using the utilization law, a form of Little's law.

This algorithm is implemented by the *Mathematica* program **sclosed** in the package **work.m**. In Chapter 4 we provide an example of the use of this model and two exercises for the reader.

Multiple Class Workload Models

Most computer systems are used simultaneously for more than one application. Some users may be entering customer orders, others developing applications, and still others may be using a spreadsheet. For multiclass models there are performance measures such as service center utilization, throughput, and response time for each individual class. This makes multiclass models more useful than single class models for most computer systems because very few computer systems can be modeled with precision as a single class model.A single class model works best for a computer system that supports only one application. For computer systems having multiple applications with substantially different characteristics, realistic modeling requires a multiclass workload model.

Although multiclass models have a number of advantages over single class models, there are a few disadvantages as well. These include:

1. A great deal more information must be collected to parameterize a multiclass model than a single class model. In some cases it may be difficult to obtain all the information needed from current measurement tools. This may lead to estimates that dilute the accuracy of the multiclass model.
2. As one would expect, multiclass model solution techniques are more difficult to implement and require more computing resources to process than single class models.

Just as with single class models, an open multiclass model is an approximation to reality but is fairly easy to implement. In Table 4.3 of Chapter 4 we outline the simple calculations necessary for the multiclass open model. This model assumes that each workload class is a transaction class. The *Mathematica* program **mopen** in the package **work.m** implements the calculations. In Chapter 4 we provide an example and an exercise that use m**open**.

The exact MVA solution algorithm for the closed multiclass model is based on the same ideas as the single class model (Little's law and the arrival theorem) but is much more difficult to explain and to implement. In addition the computational requirements have a combinatorial explosion as the number of classes and the population of each class increases. I explain the algorithm on pages 413—414 of my book [Allen 1990] and in my article [Allen and Hynes 1991] with Gary Hynes. In Chapter 4 we show how to use the *Mathematica* program **Exact** from the package **work.m,** which is a slightly revised form of the program by that name in my book [Allen 1990]. In Chapter 4 we consider some examples using **Exact**.

Unfortunately, as we mentioned earlier, **Exact** is very computationally intensive and thus is not practical for modeling systems with many workload classes or many service centers (or systems with both many workload classes and many service centers). To obviate this problem, we consider an approximate MVA algorithm for closed multiclass systems. The approximate algorithm is sufficiently accurate for most modeling studies and is much faster than the exact algorithm. We provide the *Mathematica* program **Approx** in the package **work.m** to implement the approximate algorithm; we also provide an example of its use as well as an exercise to test your understanding of the use of **Approx**.

There is an approximate MVA algorithm for modeling computer systems that (simultaneously) have both open and closed workload classes. (Recall that transaction workload classes are open although both terminal and batch workloads are closed.) The algorithm for solving mixed multiclass models is presented in my book [Allen 1990] on pages 415—416 with an example of its use. However, we do not recommend the use of this algorithm for reasons that are explained in Chapter 4.

We avoid these problems by using a modified type of closed workload class that we call a *fixed throughput* class. At the Hewlett-Packard Performance Technology Center Gary Hynes developed an algorithm that converts a terminal workload or a batch workload into a modified terminal or batch workload with a given throughput. In the case of a terminal workload we use as input the required throughput, the desired mean think time, and the service demands to create a terminal workload that has the desired throughput. We also compute the average number of active terminals required to produce the given throughput. The same algorithm works for a batch class workload because a batch workload can be thought of as a terminal workload with zero think time. For the batch class workload we compute the average number of batch jobs required to generate the required throughput.

In Chapter 4 we present an example that illustrates difficulties that arise in using transaction (open) workloads in situations in which their use seems appropriate. We also show how fixed throughput classes allow us to obtain satisfactory results. To do this we provide the *Mathematica* program **Fixed** in the package **work.m** to implement the fixed class algorithm. We also provide an exercise to test your understanding of the use of **Fixed**.

Priority Queues

In all of the models discussed so far we have assumed that there are no priorities for workload classes, that is, that all are treated the same. However, most actual computer systems do allow some workloads to have priority, that is, to receive preferential treatment over other workload classes. For example, if a computer system has two workload classes, a terminal class that is handling incoming customer telephone orders for products and the other is a batch class handling accounting or billing, it seems reasonable to give the terminal workload class priority over the batch workload class.

Every service center in a queueing network has a *queue discipline* or algorithm for determining the order in which arriving customers receive service if there is a conflict, that is, if there is more than one customer at the service center. The most common queue discipline in which there are no priority classes is the first-come, first-served assignment system, abbreviated as FCFS or FIFO (first-in, first-out). Other nonpriority queueing disciplines include last-come, first-served (LCFS or LIFO), and random-selection-for-service (RSS or SIRO).

For priority queueing systems workloads are divided into priority classes numbered from 1 to n. We assume that the lower the priority class number, the higher the priority, that is, that workloads in priority class i are given preference over workloads in priority class j if $i < j$. That is, workload 1 has the most preferential priority followed by workload 2, etc. Customers within a workload class are served with respect to that class by the FCFS queueing discipline.

There are two basic control policies to resolve the conflict when a customer of class i arrives to find a customer of class j receiving service, where $i < j$. In a *nonpreemptive* priority system, the newly arrived customer waits until the customer in service completes service before beginning service. This type of priority system is called a *head-of-the-line* system, abbreviated HOL. In a preemptive priority system, service for the priority j customer is interrupted and the newly arrived customer begins service. The customer whose service was interrupted returns to the head of the queue for the jth class. As a further refinement, in a *preemptive-resume* priority queueing system, the customer whose service was interrupted begins service at the point of interruption on the next access to the service facility.

Unfortunately, exact calculations cannot be made for networks with workload class priorities. However, widely used approximations do exist. The simplest approximation is the *reduced-work-rate* approximation for preemptive-resume priority systems that have the same priority structure at each service center. It works as follows: The processing power at node k for class c customers is reduced by the proportion of time that the service center is processing higher priority customers. Suppose the service rate of class c customers at service center k is $\mu_{c,k}$. Then the effective service rate of at node k for class c jobs is given by

$$\hat{\mu}_{c,k} = \mu_{c,k}\left(1 - \sum_{r=1}^{c-1} U_{r,k}\right).$$

The new effective service rate means that the effective service time

$$\hat{S}_{c,k} = \frac{1}{\hat{\mu}_{c,k}}.$$

Note that all customers are unaffected by lower priority customers so that, in particular, priority class 1 customers have the same effective service rate as the actual full service rate. It is also true that for class 1 workloads the network can be solved exactly.

In Chapter 4 we show how to use the reduced-work-rate approximation directly from the definition. We also show how to use the *Mathematica* program **Pri** from the package **work.m** to make the calculations and provide an exercise in the use of **Pri**.

Modeling Main Memory

Main memory is one of the most difficult computer resources to model although it is often one of the most critical resources. In many cases it must be modeled indirectly. Since the most important effect that memory has on computer performance is in its effect on concurrency, that is, allowing CPU(s), disk drives, etc., to operate independently, the most common way of modeling memory is through the multiprogramming level (MPL).

The simplest (and first) well-known queueing model of a computer system that explicitly models the multiprogramming level and thus main memory is the

Figure 8.5. Central Server Model

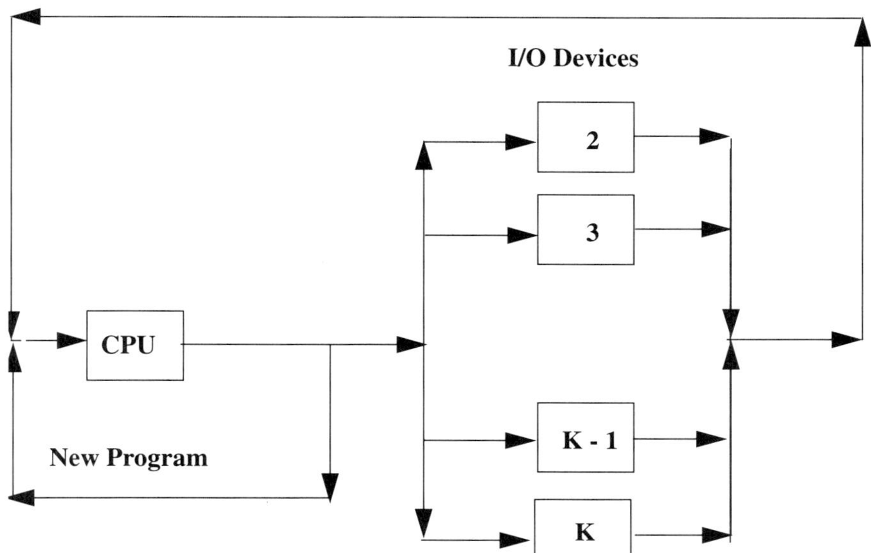

central server model shown in Figure 8.5. This model was developed by Buzen [Buzen 1971].

The central server referred to in the title of this model is the CPU. The central server model is closed because it contains a fixed number of programs N (this is also the multiprogramming level, of course). The programs can be thought of as markers or tokens that cycle around the system interminably. Each time a program makes the trip from the CPU directly back to the end of the CPU queue we assume that a program execution has been completed and a new program enters the system. Thus there must be a backlog of jobs ready to enter the computer system at all times. We assume there are K service centers with service center 1 the CPU. We assume also that the service demand at each center is known. Buzen provided an algorithm called the *convolution algorithm* to calculate the performance statistics of the central server model. In Section 4.2.4 of Chapter 4 we provide an MVA algorithm that is more intuitive and is a modification of the single class closed MVA algorithm we presented earlier in this chapter.

We provide the *Mathematica* program **cent** in the package **work.m** to implement the algorithm; in Chapter 4 we also provide examples of its use and an exercise.

Although the central server model has been used extensively it has two major flaws. The first flaw is that it models only batch workloads and only one of them at a time. That is, it cannot be used to model terminal workloads at all and it cannot be used to model more than one batch workload at a time. The other flaw is that it assumes a fixed multiprogramming level although most computer systems have a fluctuating value for this variable. In Chapter 4 we show how to adapt the central server model so that it can model a terminal or a batch workload with time varying multiprogramming level. We need only assume that there is a maximum possible multiprogramming level m.

Since a batch computer system can be viewed as a terminal system with think time zero, we imagine the closed system of Figure 8.4 as a system with N terminals or workstations all connected to a central computer system. We assume that the computer system has a fluctuating multiprogramming level with a maximum value m. If a request for service arrives at the central computer system when there are already m requests in process the request must join a queue to wait for entry into main memory. (We assume that the number of terminals, N, is larger than m.) The response time for a request is lowest when there are no other requests being processed and is largest when there are N requests either in process or queued up to enter the main memory of the central computer system. A computer system with terminals connected to a central computer with an upper limit on the multiprocessing level (the usual case) is not a BCMP queueing network. The non-BCMP model for this system is created in two steps. In the first step the entire central computer system, that is, everything but the terminals, is replaced by a *flow equivalent server* (FESC). This FESC can be thought of as a black box that when given the system workload as input responds with the same throughput and response time as the real system. The FESC is a *load dependent server*, that is, the throughput and response time at any time depends upon the number of requests in the FESC. We create the FESC by computing the throughput for the central system considered as a central server model with multiprogramming level 1, 2, 3,..., m. The second step in the modeling process is to replace the central computer system in Figure 8.4 by the FESC as shown in Figure 8.6. The algorithm to make the calculations is rather complex so we will not explain it completely here. (It is Algorithm 6.3.3 in my book [Allen 1990.) However, the *Mathematica* program **online** in the package **work.m** implements the algorithm. The inputs to **online** are m, the maximum multiprogramming level *D*emands, the vector of demands for the K service centers, N, the number of terminals, and T, the average think time. The outputs of **online** are the average throughput, the average response time, the average number of requests from the terminals that are in process, the vector of probabilities that there are 0, 1, ..., m

requests in the central computer system, the average number in the central computer system, the average time there, the average number in the queue to enter the central computer system (remember, no more than m can be there), the average time in the queue, and the vector of utilizations of the service centers.

In Example 4.9 we show how the FESC form of the central server model can be used to model the file server on a LAN.

Unfortunately, there is no easy way to extend the central server model so that it can model main memory with more than one workload class. There are expensive tools available to model memory for IBM MVS systems but they use very complex, proprietary algorithms. My colleague Gary Hynes at the Hewlett-Packard Performance Technology Center has written a modeling package that can be used to model memory for Hewlett-Packard computer systems; it is proprietary, of course.

Figure 8.6. FESC Form of Central Server Model

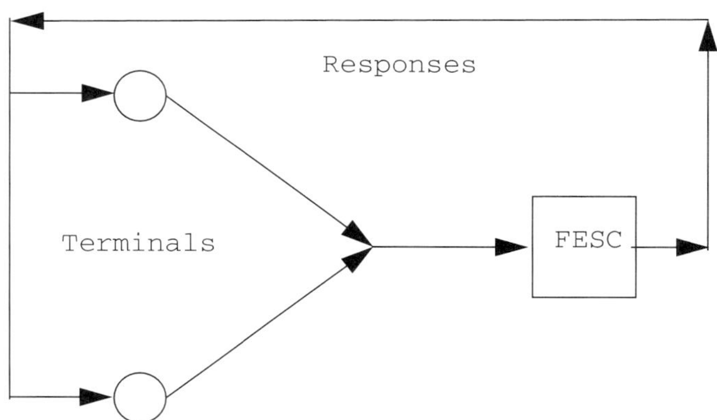

8.2.5 Chapter 5: Model Parameterization

In Chapter 5 we examine the measurement problem and the problem of parameterization. The measurement problem is, "How can I measure how well my computer system is processing the workload?" We assume that you have one or more measurement tools available for your computer system or systems. We discuss how to use your measurement tools to find out how your computer system is performing. We also discuss how to get the data you need for parameterizing a model. In many cases it is necessary to process the measurement data to obtain the parameters needed for modeling.

Monitors

The basic measurement tool for computer performance is the monitor. There are two basic types of monitors: software monitors and hardware monitors. Since Hardware monitors are used almost exclusively by computer manufacturers, we discuss only software monitors in Chapter 5. The three most common types of software monitors are used for diagnostics (sometimes called real-time or trouble shooting monitors), for studying long-term trends (sometimes called historical monitors), and job accounting monitors for gathering chargeback information. These three types can be used for monitoring the whole computer system or be specialized for a particular piece of software such as CICS, IMS, or DB2 on an IBM mainframe. There are probably more specialized monitors designed for CICS than for any other software system.

The uses for a diagnostic monitor include the following:

1. To determine the cause of poor performance at this instant.
2. To identify the user(s) and/or job(s) that are monopolizing system resources.
3. To determine why a batch job is taking an excessively long time to complete.
4. To determine whether there is a problem with the database locks.
5. To help with tuning the system.

Some diagnostic monitors have expert system capabilities to analyze the system and make recommendations to the user. A diagnostic monitor with a built-in expert system can be especially useful for an installation with no resident performance expert. An expert system or adviser can diagnose performance problems and make recommendations to the user. For example, the expert system might recommend that the priority of some jobs be changed, that the I/O load be balanced, that more main memory or a faster CPU is needed, etc. The expert system could reassure the user in some cases as well. For example, if the CPU is running at 100% utilization but all the interactive jobs have satisfactory response times and low priority batch jobs are running to fully utilize the CPU, this could be reported to the user by the expert system.

Uses for monitors designed for long term performance management include the following:

1. To archive performance data for a performance database.
2. To provide performance information needed for parameterizing models of the system.

3. To provide performance data for forecasting studies.

Most of the early performance monitors were designed to provide information for chargeback. One of the most prominent of these is the System Management Facility discussed by Merrill in [Merrill 1984] and usually referred to as SMF.

As Merrill points out, SMF information is also used for computer performance evaluation.

Accounting monitors, such as SMF, generate records at the termination of batch jobs or interactive sessions indicating the system resources consumed by the job or session. Items such as CPU seconds, I/O operations, memory residence time, etc. are recorded.

Two software monitors produced by the Hewlett-Packard Performance Technology Center are used to measure the performance of the HP-UX system I am using to write this book. HP GlancePlus/UX is an online diagnostic tool (sometimes called a trouble shooting tool) that monitors ongoing system activity. The HP GlancePlus/UX User's Manual provides a number of examples of how this monitor can be used to perform diagnostics, that is, determine the cause of a performance problem. The other software monitor used on the system is HP LaserRX/UX. This monitor is used to look into overall system behavior on an ongoing basis, that is, for trend analysis. This is important for capacity planning. It is also the tool we use to provide the information needed to parameterize a model of the system.

There are two parts of every software monitor, the collector that gathers the performance data and the presentation tools designed to present the data in a meaningful way. The presentation tools usually process the raw data to put it into a convenient form for presentation. Most early monitors were run as batch jobs and the presentation was in the form of a report, which also was generated by a batch job. While monitor collectors for long range monitors are batch jobs, most diagnostic monitors collect performance data only while the monitor is activated.

The two basic modes of operation of software monitors are called *event-driven* and *sampling*. Events indicate the start or the end of a period of activity or inactivity of a hardware or software component. For example, an event could be the beginning or end of an I/O operation, the beginning or end of a CPU burst of activity, etc. An event-driven monitor operates by detecting events. A sampling monitor operates by testing the states of a system at predetermined time intervals, such as every 10 ms.

Software monitors are very complex programs that require an intimate knowledge of both the hardware and operating system of the computer system being measured. Therefore, a software monitor is usually purchased from the computer company that produced the computer being monitored or a software performance vendor such as Candle Corporation, Boole & Babbage, Legent, Computer Associates, etc. For more detailed information on available monitors see [Howard Volume 2].

If you are buying a software monitor for obtaining the performance parameters you need for modeling your system, the properties you should look for include:

1. Low overhead.

2. The ability to measure throughput, service times, and utilization for the major servers.

3. The ability to separate workload into homogeneous classes with demand levels and response times for each.

4. The ability to report metrics for different types of classes such as interactive, batch, and transaction.

5. The ability to capture all activity on the system including system overhead by the operating system.

6. Provision of sufficient detail to detect anomalous behavior (such as a runaway process) which indicates atypical activity.

7. Provision for long term trending via low volume data.

8. Good documentation and training provided by the vendor.

9. Good tools for presenting and interpreting the measurement results.

Low overhead is important both because it leaves more capacity available for performing useful work and because high overhead distorts the measurements made by the monitor.

The problem of measuring system CPU overhead has always been a challenge at IBM MVS installations. It is often handled by "capture ratios." The capture ratio of a job is the percentage of the total CPU time for a job that has been captured by SMF and assigned to the job. The total CPU time consists of the TCB (task control block) time plus the SRB (system control block) time plus the overhead, which normally cannot be measured. It may require some less than straightforward calculations to convert the measured values of TCB and SRB provided by SMF records into actual times in seconds. For an example of these calculations see [Bronner 1983]. For an overview of RMF see [IBM 1991]. If the capture ratio for a job or workload class is known, the total CPU utilization can be obtained by dividing the sum of the TCB time and the SRB time by the capture ratio. The CPU capture ratio can be estimated by linear regression and other techniques. Wicks describes how to use the regression technique in Appendix D of [Wicks 1991]. The approximate values of the capture ratio for many types of applications are known. For example, for CICS it is usually between 0.85 and 0.9, for TSO between 0.35 and 0.45, for commercial batch workload classes between 0.55 and 0.65, and for scientific batch workload classes between 0.8 and 0.9.

We illustrate the calculation of capture ratios in Example 5.1.

We provide a further discussion of the modeling study paradigm in Section 5.3.1. (We had discussed it earlier in Section 3.5.)

8.2.6 Chapter 6: Simulation and Benchmarking

Simulation and benchmarking have a great deal in common. That is why Hamming [Hamming 1991] said, "Simulation is better than reality!" When simulating a computer system we manipulate a model of the system; when benchmarking a computer system we manipulate the computer system itself. Manipulating the real computer system is more difficult and much less flexible than manipulating a simulation model. In the first place, we must have physical possession of the computer system we are benchmarking. This usually means it cannot be doing any other work while we are conducting our benchmarking studies. If we find that a more powerful system is needed, we must obtain access to the more powerful system before we can conduct benchmarking studies on it. By contrast, if we are dealing with a simulation model, in many cases, all we need to do to change the model is to change some of the parameters.

For benchmarking an online system, in most cases, part of the benchmarking process is simulating the online input used to drive the benchmarked system. This is called "remote terminal emulation" and usually is performed on a second computer system, which transmits the simulated online workload to the computer under study. The simulator that performs the remote terminal emulation is called a *driver*. Remote terminal emulation is the method most commonly used to simulate the online workload classes. Thus simulation modeling is also part of benchmark modeling for most benchmarks that include terminal workloads.

Another common feature of simulation and benchmarking is that a simulation run and a benchmarking run are both examples of a random process and thus must be analyzed using statistical analysis tools. The proper analysis of simulation output and benchmarking output is a key part of simulation or benchmarking; such a study without proper analysis can lead to the wrong conclusions.

Simulation

The kind of simulation that is most important for modeling computer systems is often called *discrete event simulation* but certainly falls within the rubric of what Knuth calls the Monte Carlo method. Knuth in his widely referenced book [Knuth 1981], says, "These traditional uses of random numbers have suggested the name 'Monte Carlo method,' a general term used to describe any algorithm that employs random numbers."

Twenty years ago modeling computer systems was almost synonymous with simulation. Since that time so much progress has been made in analytic queueing theory models of computer systems that simulation has been displaced by queueing theory as the modeling technique of choice; simulation is now considered by many computer performance analysts to be the modeling technique of last resort. Most modelers use analytic queueing theory if possible and simulation only if it is very difficult or impossible to use queueing theory. Most current computer system modeling packages use queueing network models that are solved analytically.

The reason for the preference by most analysts for analytic queueing theory modeling is that it is much easier to formulate the model and takes much less computer time to use than simulation. See, for example, the paper [Calaway 1991] we discussed in Chapter 1.

When using simulation as the modeling tool for a modeling study the first step of the modeling study paradigm discussed in Section 5.3.1 is especially important, that is, to define the purpose of the modeling study.

Bratley, Fox, and Schrage [Bratley, Fox, and Schrage 1987] define simulation as follows:

> Simulation means driving a model of a system with suitable inputs and observing the corresponding outputs.

Thus simulation modeling is a process that is much like measurement of an actual system. It is essentially an experimental procedure. In simulation we mimic or emulate an actual system by running a computer program (the simulation model) that behaves much like the system being modeled. We predict the behavior of the actual system by measurements made while running the simulation model. The simulation model generates customers (workload requests) and routes them through the model in the same way that a real workload moves through a computer system. Thus visits are made to a representation of the CPU, representations of I/O devices, etc.

To perform steps 4 and 5 of the modeling study paradigm described in Section 5.3.1 (and more briefly in Section 3.5) requires the following basic tasks.

1. Construct the model by choosing the service centers, the service center service time distributions, and the interconnection of the center.
2. Generate the transactions (customers) and route them through the model to represent the system.
3. Keep track of how long each transaction spends at each service center. The service time distribution is used to generate these times.
4. Construct the performance statistics from the above counts.
5. Analyze the statistics.
6. Validate the model.

Of course, these same tasks are necessary for Step 6 of the modeling study paradigm.

Simulation is a powerful modeling techniques but requires a great deal of effort to perform successfully. It is much more difficult to conduct a successful modeling study using simulation than is generally believed.

Challenges of modeling a computer system using simulation include:

1. Determining the goal of the study.
2. Determining whether or not simulation is appropriate for making the study. If so, determine the level of detail required. It is important to schedule sufficient time for the study.
3. Collecting the information needed for conducting the simulation study. Information is needed for validation as well as construction of the model.
4. Choosing the simulation language. This choice depends upon the skills of the people available to do the coding.
5. Coding the simulation, including generating the random number streams needed, testing the random number streams, and verifying that the coding is correct. People with special skills are needed for this step.

6. Overcoming the special simulation problems of determining when the simulation process has reached the steady-state and a method of judging the accuracy of the results.

7. Validating the simulation model.

8. Evaluating the results of the simulation model.

A failure of any one of these steps can cause a failure of the whole effort.

We discuss all of these simulation challenges with examples and exercises in Chapter 6.

Benchmarking

There are actually two basically different kinds of benchmarking. The first kind is defined by Dongarra *et al.* [Dongarra, Martin, and Worlton 1987] as "Running a set of well-known programs on a machine to compare its performance with that of others." Every computer manufacturer runs these kinds of benchmarks and reports the results for each announced computer system. The second kind is defined by Artis and Domanski [Artis and Domanski 1988] as "a carefully designed and structured experiment that is designed to evaluate the characteristics of a system or subsystem to perform a specific task or tasks." The first kind of benchmark is represented by the Whetstone, Dhrystone, and Linpack benchmarks.

The Artis and Domanski kind of benchmark is the type one would use to model the workload on your current system and run on the proposed system. It is the most difficult kind of modeling in current use for computer systems.

Before we discuss the Artis and Domanski type of benchmark we discuss the first type of benchmark, the kind that is called a *standard* benchmark.

The two best known standard benchmarks are the Whetstone and the Dhrystone. The Whetstone benchmark was developed at the National Physical Laboratory in Whetstone, England, by Curnow and Wichman in 1976. It was designed to measure the speed of numerical computation and floating-point operations for midsize and small computers. Now it is most often used to rate the floating-point operation of scientific workstations. My IBM PC compatible 33 MHz 486 has a Whetstone rating of 5,700K Whetstones per second. According to [Serlin 1986] the HP3000/930 has a rating of 2,841K Whetstones per second, the IBM 4381-11 has a rating of approximately 2,000K Whetstones per second, and the IBM RT PC a rating of 200K Whetstones per second.

The Dhrystone benchmark was developed by Weicker in 1984 to measure the performance of system programming types of operating systems, compilers, editors, etc. The result of running the Dhrystone benchmark is reported in Dhrystones per second. Weicker in his paper [Weicker 1990] describes his original benchmark as well as Versions 1.1 and 2.0. Weicker [Weicker 1990] not only discusses his Dhrystone benchmark but also discusses the Whetstone, Livermore Fortran Kernels, Stanford Small Programs Benchmark Set, EDN Benchmarks, Sieve of Eratosthenes, and SPEC benchmarks. Weickert's paper is one of the best summary papers available on standard benchmarks.

According to QAPLUS Version 3.12, my IBM PC 33 MHz 486 compatible executes 22,758 Dhrystones per second. According to [Serlin 1986] the IBM 3090/200 executes 31,250 Dhrystones per second, the HP3000/930 executes 10,000 Dhrystones per second, and the DEC VAX 11/780 executes 1,640 Dhrystones per second, with all figures based on the Version 1.1 benchmark. However, IBM calculates VAX MIPS by dividing the Dhrystones per second from the Dhrystone 1.1 benchmark by 1,757; IBM evidently feels that the VAX 11/780 is a 1,757 Dhrystones per second machine. The Dhrystone statistics on the 11/780 are very sensitive to the version of the compiler in use. Weicker [Weicker 1990] reports that he obtained very different results running the Dhrystone benchmark on a VAX 11/780 with Berkeley UNIX (4.2) Pascal and with DEC VMS Pascal (V.2.4). On the first run he obtained a rating of 0.69 native MIPS and on the second run a rating of 0.42 native MIPS. He did not reveal the Dhrystone ratings.

Standard benchmarks are useful in providing at least ballpark estimates of the capacity of different computer systems. However there are a number of problems with the older standard benchmarks such as Whetstone, Dhrystone, Linpack, etc. One problem is that there are a number of different versions of these benchmarks and vendors sometimes fail to mention which version was used. In addition, not all vendors execute them in exactly the same way. That is apparently the reason why Checkit, QAPLUS, and Power Meter report different values for the Whetstone and Dhrystone benchmarks. Another complicating factor is the environment in which the benchmark is run. These could include operating system version, compiler version, memory speed, I/O devices, etc. Unless these are spelled out in detail it is difficult to interpret the results of a standard benchmark.

Three new organizations have been formed recently with the goal of providing more meaningful benchmarks for comparing the capability of computer systems for doing different types of work. The Transaction Processing Performance Council (TPC) was founded in 1988 at the initiative of Omri Serlin to develop online teleprocessing (OLTP) benchmarks. Just as the TPC was organized to develop benchmarks for OLTP the Standard Performance Evaluation Corporation (SPEC) is a nonprofit corporation formed to establish, maintain, and endorse a standardized set of benchmarks that can be applied to the newest generation of high-performance computers and to assure that these benchmarks are consistent and available to manufacturers and users of high-performance systems. The four founding members of SPEC were Apollo Computer, Hewlett-Packard, MIPS Computer Systems, and Sun Microsystems. The Business Applications Performance Corporation (BAPCo) was formed in May 1991. It is a nonprofit corporation that was founded to create for the personal computer user objective performance benchmarks that are representative of the typical business environment. Members of BAPCo include Advanced Micro Devices Inc., Digital Equipment, Dell Computer, Hewlett-Packard, IBM, Intel, Microsoft, and Ziff-Davis Labs.

In Chapter 6 we discuss the benchmarks developed by SPEC, TPC, and BAPCo and present some representative results of these benchmarks.

Drivers (RTEs)

To perform some of the benchmarks we mention in Chapter 6, such as the TPC benchmarks TPC-A and TPC-C, a special form of simulator called a *driver* or *remote terminal emulator* (RTE) is used to generate the online component of the workload. The driver simulates the work of the people at the terminals or workstations connected to the system as well as the communication equipment and the actual input requests to the computer system under test (SUT in benchmarking terminology). An RTE, as shown in Figure 8.7, consists of a separate computer with special software that accepts configuration information and executes job scripts to represent the users and thus generate the traffic to the SUT. There are communication lines to connect the driver to the SUT. To the SUT the input is exactly the same as if real users were submitting work from their terminals. The benchmark program and the support software such as compilers or database management software are loaded into the SUT, and driver scripts representing the users are placed on the RTE system. The RTE software reads the scripts, generates requests for service, transmits the requests over the communication lines to the benchmark on the SUT, waits for and times the responses from the benchmark program, and logs the functional and performance information. Most drivers also have software for recording a great deal of statistical performance information.

Most RTEs have two powerful software features for dealing with scripts. The first is the ability to capture scripts from work as it is being performed. The second is the ability to generate scripts by writing them out in the format understood by the software.

Figure 8.7. A Remote Terminal Emulator (RTE)

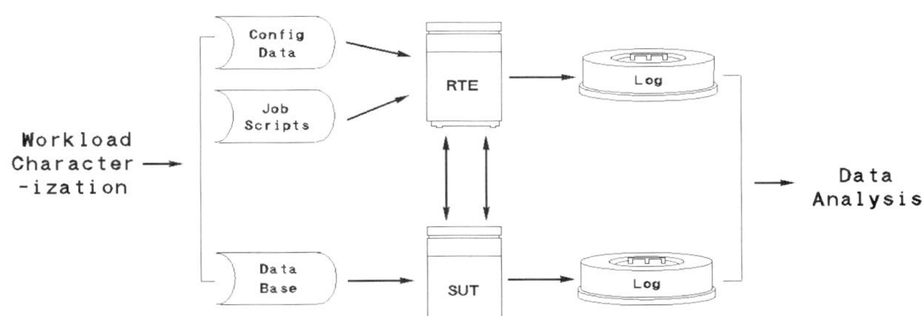

All computer vendors have drivers for controlling their benchmarks. Since there are more IBM installations than any other kind, the IBM Teleprocessing Network Simulator (program number 5662-262, usually called TPNS) is probably the best known driver in use. TPNS generates actual messages in the IBM Communications Controller and sends them over physical communication lines (one for each line that TPNS is emulating) to the computer system under test.

TPNS consists of two software components, one of which runs in the IBM mainframe or plug compatible used for controlling the benchmark and one that runs in the IBM Communications Controller. TPNS can simulate a specified network of terminals and their associated messages, with the capability of altering network conditions and loads during the run. It enables user programs to operate as they would under actual conditions, since TPNS does not simulate or affect any functions of the host system(s) being tested. Thus it (and most other similar drivers including WRANGLER, the driver used at the Hewlett-Packard Performance Technology Center) can be used to model system performance, evaluate communication network design, and test new application programs. A driver may be much less difficult to use than the development of some detailed simulation models but is expensive in terms of the hardware required. One of its most important uses is testing new or modified online programs both for accuracy and performance. Drivers such as TPNS or WRANGLER make it possible to utilize all seven of the uses of benchmarks described by Artis and Domanski. Kube [Kube 1981] describes how TPNS has been used for all these activities. Of course the same claim can be made for most commercial drivers.

Developing Your Own Benchmark for Capacity Planning

Unless your objectives are very limited or your workload is very simple, developing your own benchmark for predicting future performance on your current system or an upgraded system is rather daunting. By "predicting future performance" we mean predicting performance with the workload you forecast for the future. Experienced benchmark developers complain about "the R word," that is, developing a benchmark that is truly representative of your actual or future workload.

In spite of all the difficulties and challenges we discuss in Chapter 6, it is possible to construct representative and useful benchmarks. Computer manufacturers couldn't live without them and some large computer installations depend upon them. However, constructing a good benchmark for your installation is not an easy task and is not recommended for most installations. Incorvia [Incorvia 1992] examines benchmark costs, risks, and alternatives for mainframe computers. He concludes with the following recommendations:

> Before your staff initiates plans to develop a benchmark, collect all available performance information on mainframes you are evaluating. Include the sources noted here, and any other sources which you feel are reasonable.
>
> Take sufficient time to produce, review, and distribute a formal report of your findings. *After* the review process, determine the incremental value involved in doing a benchmark. If there is insufficient incremental value to justify a quality benchmark, *don't do one*.
>
> Alternatively, develop a representative, natural ETR-based, externally driven benchmark. This is the benchmark we've discussed with costs between $600,000 and $1 million. If you plan to do this, allow one year lead time. You will also need significant executive management commitment, start-up budget, education, stand-alone time, and budget for significant recurring costs.
>
> If you decide to develop a high quality benchmark, contact your suppliers early in the development cycle. Suppliers have considerable experience in the development of such benchmarks, and will be eager to assist you and corroborate their benchmark results.

8.2.7 Chapter 7: Forecasting

Forecasting is the technique for performance management that is most familiar to business people not in IS. Almost every business uses forecasting for some purposes. Time series analysis is one of the most prevalent forecasting techniques. Forecasting using time series analysis is essentially a form of pattern recognition or curve fitting. The most popular pattern is a straight line but other patterns sometimes used include exponential curves and the S-curve. One of the keys to good forecasting is good data and the source of much useful data is the user community. That is why one of the most popular and successful forecasting techniques for computer systems performance management is forecasting using natural forecasting units (NFUs), also known as business units (BUs) and as key volume indicators (KVI). The users can forecast the growth of natural forecasting units such as new checking accounts, new home equity loans, or new life insurance policies much more accurately than computer capacity planners in the installation can predict future computer resource requirements from past requirements. If the capacity planners can associate the computer resource usage with the natural forecasting units, future computer resource requirements can be predicted. For example, it may be true that the CPU utilization for a computer system is strongly correlated with the number of new life insurance policies sold by the insurance company. Then, from the predictions of the growth of policies sold, the capacity planning group can predict when the CPU utilization will exceed the threshold requiring an upgrade.

NFU Time Series Forecasting

NFU forecasting is a form of time series forecasting. Time series forecasting is a discipline that has been used for applications such as studying the stock market, the economic performance of a nation, population trends, rainfall, and many others. An example of a time series that we might study as a computer performance analyst is $u_1, u_2, u_3, \ldots, u_n, \ldots$ where u_i is the maximum CPU utilization on day i for a particular computer system.

All the major statistical analysis systems such as SAS and Minitab provide tools for the often complex calculations that go with time series analysis. For the convenience of computer performance analysts who have Hewlett-Packard computer equipment the Hewlett-Packard Performance Technology Center has developed HP RXForecast for HP 3000 MPE/iX computer systems and for HP 9000 HP-UX computer systems.

Natural forecasting units are sometimes called business units or key volume indicators because an NFU is usually a business unit. The papers [Browning 1990], [Bowerman 1987], [Reyland 1987], [Lo and Elias 1986], and [Yen 1985] are some of the papers on NFU (business unit) forecasting that have been presented at National CMG Conferences. In their paper [Lo and Elias 1986], Lo and Elias list a number of other good NFU forecasting papers.

The basic problem that NFU forecasting solves is that the end users, the people who depend upon computers to get their work done, are not familiar with computer performance units (sometimes called DPUs for data processing units) such as interactions per second, CPU utilization, or I/Os per second, while computer capacity planners are not familiar with the NFUs or the load that NFUs put on a computer system.

Lo and Elias [Lo and Elias 1986] describe a pilot project undertaken at their installation. According to Lo and Elias, the major steps needed for applying the NFU forecasting technique are (I have changed the wording slightly from their statement):

1. Identify business elements as possible NFUs.
2. Collect data on the NFUs.
3. Determine the DPUs of interest.
4. Collect the DPU data.
5. Perform the NFU/DPU dependency analysis.
6. Forecast the DPUs from the NFUs.
7. Determined the capacity requirement from the forecasts.
8. Perform an iterative review and revision.

Lo and Elias used the Boole & Babbage Workload Planner software to do the dependency analysis. This software was also used to project the future capacity requirements using standard linear and compound regression techniques.

Yen, in his excellent paper [Yen 1985], describes how he predicted future CPU requirements for his IBM mainframe computer installation from input from users. He describes the procedure in the abstract for his paper as follows:

> Projecting CPU requirements is a difficult task for users. However, projecting DASD requirements is usually an easier task.

> This paper describes a study which demonstrates that there is a positive relationship between CPU power and DASD allocations, and that if a company maintains a consistent utilization of computer processing, it is possible to obtain CPU projections by translating users DASD requirements.

Yen discovered that user departments can accurately predict their magnetic disk requirements (IBM refers to magnetic disks as DASD for "direct access storage device"). They can do this because application developers know the record sizes of files they are designing and the people who will be using the systems can make good predictions of business volumes. Yen used 5 years of historical data describing DASD allocations and CPU consumption in a regression study. He made a scatter diagram in which the y-axis represented CPU hours required for a month, Monday through Friday, 8 am to 4 pm, while the x-axis represented GB of DASD storage installed online on the fifteenth day of that month. Yen found that the regression line $y = 34.58 + 2.59x$ fit the data extraordinarily well. The usual measure of goodness-of-fit is the R-squared value, which was 0.95575. (R-squared is also called the *coefficient of determination.*) In regression analysis studies, R-squared can vary between 0, which means no correlation between x and y values, and ,1 which means perfect correlation between x and y values. A statistician might describe the R-squared value of 0.95575 by saying, "95.575 percent of the total variation in the sample is due to the linear association between the variables x and y." An R-squared value larger than 0.9 means that there is a strong linear relationship between x and y.

Yen was able to make use of his regression equation plus input from some application development projects to predict when the next computer upgrade was needed.

Yen no longer has the data he used in his paper but provided me with data from December 1985 through October 1990. From this data I obtained the x and y values plotted in Figure 8.8 together with the regression line obtained using the package **LinearRegression** from the **Statistics** directory of *Mathematica*. The x values are GB of DASD storage online as of the fifteenth of the month, while y is the measured number of CPU hours for the month, normalized into 19 days of 8 hours per day measured in units of IBM System/370 Model 3083 J processors. The Parameter Table in the output from the **Regress** program shows that the regression line is $y = -310.585 + 2.25101\ x$, where x is the number of GB of online DASD storage and y is the corresponding number of CPU hours for the month. We also see that R-squared is 0.918196 and that the estimates of the constants in the regression equation are both considered significant. If you are well versed in statistics you know what the last statement means. If not, I can tell you that it means that the estimates look very good. Further information is provided in the ANOVA Table to bolster the belief that the regression line fits the data very well. However, a glance at Figure 8.8 indicates there are several points in the scatter diagram that appear to be outliers. (An outlier is a data point that doesn't seem to belong to the remainder of the set.) Yen has assured me that the two most prominent points that appear to be outliers really are! The leftmost outlier is the December 1987 value. It is the low point just above the x-axis at $x = 376.6$. Yen says that the installation had just upgraded their DASD so that there was a big jump in installed online DASD storage. In addition, Yen recommends taking out all December points because every December is distorted by extra holidays. The rightmost outlier is the point for December 1989, which is located at (551.25, 627.583). Yen says the three following months are outliers as well, although they don't appear to be so in the figure. Again, the reason these points are outliers is another DASD upgrade and file conversion.

```
In[3]:= <<Statistics'LinearRegression'
In[12]:= Regress[data, {1,x}, x]

Out[12]= {ParameterTable ->
    Estimate    SE           TStat        PValue,
 1   -310.585   34.1694      -9.08955    0

 x   2.25101    0.0889939    25.294      0

>    RSquared -> 0.918196, AdjustedRSquared ->
0.91676,

>    EstimatedVariance -> 3684.01,

>    ANOVATable ->
      DoF   SoS              MeanSS           FRatio
PValue}

Model    1     2.35697 10     2.35697 10     639.785    0
Error    57    209989.        3684.01
                                6
     Total    58    2.56696 10
```

Figure 8.8. Scatter Diagram For Yen Data

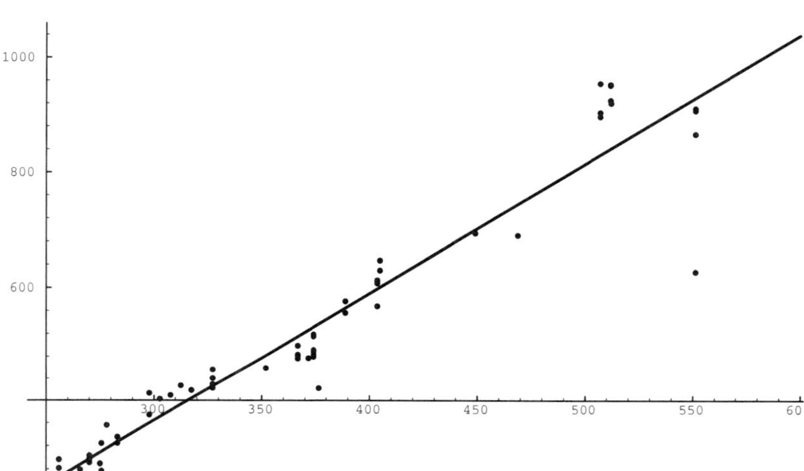

Here we show the Parameter Table from **regress** with the outliers removed.

```
Out[7]= {ParameterTable ->

    Estimate    SE            TStat        PValue,
 1    -385.176    25.6041       -15.0435    0

 x    2.48865     0.0688442    36.149       0

>     RSquared -> 0.963858, AdjustedRSquared ->
0.96312,

>     EstimatedVariance -> 1478.93,

 >     ANOVATable ->
        DoF    SoS               MeanSS         FRatio
PValue}
                              6              6
 Model    1      1.9326 10      1.9326 10     1306.75    0
Error    49    72467.7          1478.93
                                6
 Total    50     2.00507 10
```

All of the statistical tables got a little scrambled by the capture routine. However, the results are now definitely improved with R-squared equal to 0.963858 and the regression line $y = -385.176 + 2.48865\ x$. The new plot in Figure 8.9 clearly shows the improvement.

Figure 8.9. Regression Line For Corrected Data

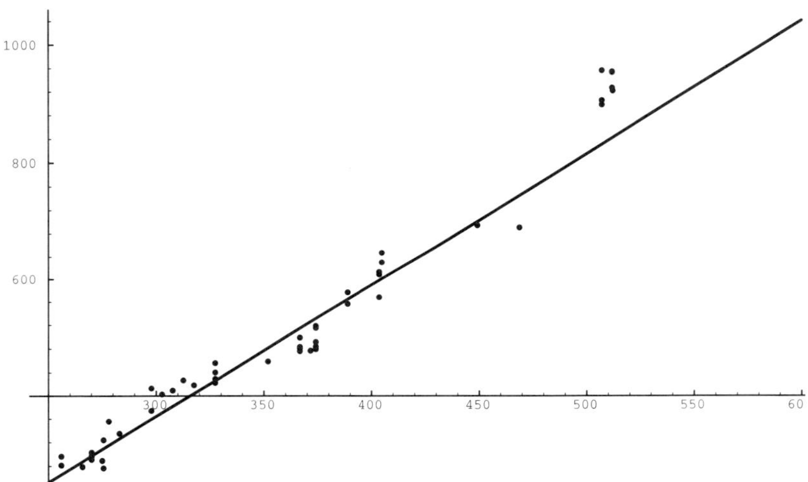

In Chapter7 we provide an example (Example 7.2) of workload forecasting taken from the HP RXForecast User's Manual. HP RXForecast was used to correlate the global CPU utilization to the business units provided in the business unit file TAHOEWK.BUS. From this information RXForecast was able to predict the global CPU utilization from the predicted business unit as shown in Figure 7.3, reproduced here as Figure 8.10. Note that for this technique to work, the predicted growth of business units must be provided to HP RXForecast.

Figure 8.10. Business Unit Forecasting Example

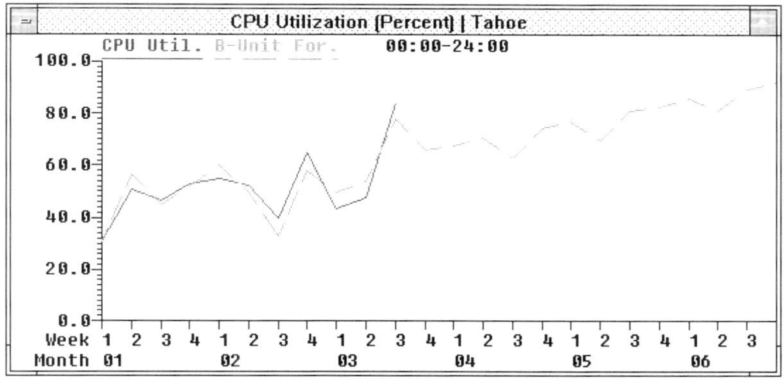

What we have here is a failure to communicate.
Warden to Paul Newman
Cool Hand Luke

8.3 Recommendations

This book is an introductory one so that, even if you have absorbed every word in it, there is still much to be learned about computer performance management. In this section I make recommendations about how to learn more about performance management of computer systems from both the management and purely technical views. There is much more material available on the technical side than the management side. In fact, I have not been able to find even one outstanding contemporary book on managing computer performance activities. The book [Martin *et al.* 1991] is an excellent book on the management of IS that emphasizes the importance of good performance but provides little information on how to achieve good performance. In spite of this weakness, if you are part of IS management, you should read this book. It provides a number of good references, an excellent elementary introduction to computer systems as well as telecommunications and networking, and sections on all aspects of IS management. Another useful but brief book [Lam and Chan 1987] discusses capacity planning from a management point of view. It features the results of an empirical study of computer capacity planning practices based on a survey the authors made of the 1985 Fortune 1000 companies. Lam and Chan base their conclusions on the 388 responses received to their questionnaire. (They mailed 930 questionnaires; 388 usable replies were returned.) The Lam and Chan book also has an excellent bibliography with both management and technical references.

Neither of these books covers in detail some of the most important management tools such as service level agreements, chargeback, and software performance engineering. (The brief book [Dithmar, Hugo, and Knight 1989] provides a lucid discussion of service level agreements with an excellent example service level agreement with notes.) The best source for written information on these techniques is the collection of articles mentioned in Chapter 1 and listed in the references to that chapter. A few are listed at the end of this chapter as well. (The papers on service level agreements [Miller 1987a] and [Duncombe 1992] are especially recommended.) These should be supplemented with articles published by the Computer Measurement Group in the annual proceedings for the December meeting and in their quarterly publication *CMG Reviews*. (The paper by Rosenberg [Rosenberg 1992] is highly recommended both for its wisdom and its entertaining style.) Another source of good management articles is *The Capacity Management Review* published by the Institute for Computer Capacity Management. This organization also publishes six volumes of their *IS Capacity Management Handbook Series*, which is updated on a regular basis and contains a great deal of information that is valuable for managers of computer installations. The institute also publishes technical reports such as their 1989 report *Managing Customer Service*.

If you are going to implement a new technique such as the negotiation of service level agreements with your users, the implementation of a chargeback system, or both techniques, the most efficient way to learn how to do so without excessive pain is to attend a class or workshop on each such technique. If you work for a company that uses techniques such as service level agreements and chargeback, there are probably classes or workshops available, internally. If not, the Institute for Computer Capacity Management has the following courses or workshops that could be of help: Costing and Chargeback Workshop, Managing IS Costs, and Managing Customer Service. [Of the 13 organizations I have identified that provide training in performance management related areas, only the Institute for Computer Capacity Management (ICCM) offers instruction in service level management and chargeback except, possibly, as part of a more general course.] If you are contemplating starting a capacity planning program, there are even more training opportunities including the following: Introduction to IS Capacity Management (ICCM), Preparing a Formal Capacity Plan (ICCM), Basic Capacity Planning (Watson and Walker, Inc.), and Capacity Planning (Hitachi Data Systems).

One important area of performance management that we were unable to include in this book is the general area of computer communication networks. The most important application of these networks is *client/server computing*, sometimes called distributed processing, cooperative processing, or even transaction processing and described as "The network is the computer." I describe it in [Allen 1993]: "Client/server computing refers to an architecture in which applications on intelligent workstations work transparently with data and applications on other processors, or servers, across a local or wide area network." To understand client/server computing you must, of course, understand computer communication networks. A very simple nontechnical introduction to such networks is provided in Chapter 6 of [Martin *et al.* 1991]. For a more detailed, technical description that is very clearly written see [Tanenbaum 1988]. (Tanenbaum's book comes close to being the standard computer network book for technical readers.) A more elementary discussion is provided by [Miller 1991]. I wrote a tutorial [Allen 1993] about client/server computing. There are a number of technical books about the subject including [Berson 1992], [Inmon 1993], and [Orfali and Harkey 1992]. The book by Inmon is the least technical of these books but very clearly written and highly recommended. Although we do not discuss computer communication networks or client/server computing in this book, many of the tools we discussed are valuable in studying the performance of these systems. For example, in their paper [Turner, Neuse, and Goldgar 1992], Turner *et al.* discuss how to use simulation to study the performance of a client/server system. Similarly, Swink [Swink 1992] shows how SPE can be utilized in the client/server environment.

A number of computer communication network short courses (2 to 5 days) are taught by the following vendors: QED Information Sciences, Amdahl Education, Data-Tech Institute, and Technology Exchange Company. There are also a number of client/server courses including: Building Client/Server Applications (Technology Training Corp.), How to Integrate Client-Server Into the IBM Environment (Technology Transfer Institute), Managing the Migration to Client-Server Architectures (Microsoft University), Analysis and Design of Client-Server Systems (Microsoft University), and Implementing Client/Server Applications and Distributed Data (Digital Consulting, Inc.).

To learn more about the components of computer performance, the subject of Chapter 2, you may want to read the outstanding book [Hennessy and Patterson 1990].

As Lam and Chan mention in Chapter 3 of [Lam and Chan 1987, two basic modeling approaches are used for modeling computer systems for the purpose of performance management—the component approach and the systems approach. We used the systems approach when we used queueing network models in Chapter 4 but many small installations as well as some very large installations use the components approach. Lam and Chan describe this approach as follows:

> The underlying concept of this approach is that each component in a computer system is treated largely as an independent unit, including the CPU, memory, I/O channels, disks, printers, etc. The capacity of the CPU, for example, is usually defined as the utilizable CPU hours available per day, per week, per month, etc., taking into account the hours of operation, scheduled maintenance, unscheduled system down time due to hardware or software failures, reruns due to human or machine errors, capacity limit rules of thumb, and so forth.

Installations that take this approach tend to use very simple modeling techniques such as rules-of-thumb. Others use more sophisticated techniques such as queueing theory or simulation but apply them to the component of the system most likely to be the bottleneck such as the CPU or an I/O device. Very simple queueing theory models can sometimes be applied to components. By simple we mean an open queueing system with a single service center. Queueing theory was originally developed for the study of telephone systems using simple but powerful models. These same models have been used to study I/O devices including channels and disks, caches, and LANs. My book [Allen 1990] covers these simple queueing models as well as the more complex queueing network models used in Chapter 4 of this book. My self-study course [Allen 1992] uses my book as a textbook and includes a modeling package that runs under Microsoft Windows 3.x. The two volumes [Kleinrock 1975, Kleinrock 1976] are the definitive books on queueing theory; they are praised by theoreticians as well as practitioners and cover most aspects of the theory as it applies to computer systems and networks.

The elegant and elementary book [Hall 1991] is especially recommended for learning beginning queueing theory, although none of the examples in the book concern computer system performance. The book has an excellent chapter on simulation as well as a number of examples of the use of simulation throughout the book. In addition there is an outstanding chapter on queue discipline and many examples of how to improve the performance of a queueing system including a chapter on how to design a system in which people must be subjected to some queueing (waiting). Hall says the concerns are:

1. Creating a pleasant waiting environment.
2. Implementing effective and appropriate queue disciplines.
3. Planning a queueing layout that promotes ease of movement and avoids crowding.
4. Locating servers so that they are convenient to customers, while minimizing waiting
5. Providing sufficient space to accommodate ordinary queue sizes.

Hall closes this chapter as follows:

> The message of this chapter is that actions can be taken to alleviate the consequences of queueing. Queueing need not be unpleasant. Queueing need not be chaotic. But no matter what, queueing should be prevented. It should be prevented because it takes away the customer's freedom to do as he or she chooses. Nevertheless, after all avenues for eliminating queues have been exhausted, occasional queueing might still remain. The last step is then to design the queue—to create a pleasant environment capable of accommodating ordinary queue sizes.

Don't you wish Professor Hall had designed your computer room or the waiting room of your HMO? It would be difficult to praise Randolph's book too highly!

The standard book on the use of analytic queueing theory network models to study the performance of computer systems using MVA (Mean Value Analysis) is [Lazowska *et al.* 1984]. More recent books on the subject include [King 1990], [Molloy 1989] and [Robertazzi 1990]. Computer installations that use analytic queueing theory network models often find that it is more cost effective to purchase a modeling package than to develop the software required to make the calculations. Most available modeling packages are described in [Howard Volume 1]. Vendors for the software also provide the training necessary to use the products.

A number of good books are available on simulation, and simulation is taught at many universities. In addition, authors of simulation books sometimes offer simulation courses at the extension divisions of universities. Thus, it is not terribly difficult to learn the basics of simulation. However, there are special problems with simulating computer systems so that books or papers that provide solutions for these problems are especially valuable for computer performance studies that use simulation. My favorite paper on simulation (actually, it is a chapter of a book) is [Welch 1983]. If you have any interest in simulation, especially as it applies to computer system performance modeling, you should read Welch's paper. Welch's paper appears in [Lavenberg 1983], a book that contains several other excellent chapters on simulation as well as analytic queueing theory. Another good reference for simulation modeling is the April 1981 issue of *Communications of the ACM*, which is a special issue on simulation modeling and statistical computing.

While there are general purpose simulation packages such as SIMSCRIPT II.5 that can be used to model computer systems, it is usually easier to use simulation modeling packages that were explicitly designed for modeling computer systems. A number of these are described in [Howard Volume 1]. A typical example of such a system is PAWS-PERFORMANCE ANALYST WORKBENCH SYSTEM. According to the description in [Howard Volume 1]:

> PAWS is a computer performance modeling language for the performance-oriented design of new systems as well as the analysis of existing systems.... The PAWS model definition language contains high-level computer oriented primitives such as interrupts, forks and joins, processor scheduling disciplines, and passive resources (for modeling peripheral processors, channels, buffers, control points, etc.), which allow the user to incorporate a primitive simply by specifying its name....

Other simulation modeling packages designed for modeling computer systems, of course, have similar capabilities. Vendors of such packages normally provide training for their customers.

Benchmarking is the most difficult modeling approach to learn. The book [Ferrari, Serazzi, and Zeigner 1983] is an excellent book and contains an introduction to benchmarking but was written before some of the important recent developments in benchmarking occurred, such as the founding of the TPC and SPEC organizations. Very few classes are taught on benchmarking so one has to learn by reading articles such as [Morse 1993] and [Incorvia 1992] and by serving an apprenticeship under an expert. There is no royal road to benchmarking.

Forecasting is a discipline that is widely used by management, is well documented in books and articles, and is taught not only in colleges and universities but also by those who offer training in computer performance management. In addition, there are a number of workload forecasting tools available and listed in [Howard Volume 1].

I hope you have found this book useful. If you have questions or suggestions for the second edition, please write to me; if it is extremely urgent, call me. My address is: Dr. Arnold Allen, Hewlett-Packard, 8000 Foothills Boulevard., Roseville, CA 95747. My phone number is (916) 785-5230.

8.4 References

1. Arnold O. Allen, *Probability, Statistics, and Queueing Theory with Computer Science Applications,* Second Edition, Academic Press, San Diego, 1990.

2. Arnold O. Allen, "So you want to communicate? Can open systems and the client/server model help?," *Capacity Planning and Alternative Platforms*, Institute for Computer Capacity Management, 1993.

3. Arnold O. Allen and Gary Hynes, "Approximate MVA solutions with fixed throughput classes," *CMG Transactions* (71), Winter 1991, 29—37.

4. Arnold O. Allen and Gary Hynes, "Solving a queueing model with Mathematica," *Mathematica Journal*, 1(3), Winter 1991, 108—112.

5. H. Pat Artis and Bernard Domanski, *Benchmarking MVS Systems*, Notes from the course taught January 11—14, 1988, at Tyson Corner, VA.

6. Forest Baskett, K. Mani Chandy, Richard R. Muntz, and Fernando G. Palacios, "Open, closed, and mixed networks of queues with different classes of customers," *JACM*, 22(2), April 1975, 248—260.

7. Alex Berson, *Client/Server Architecture,* McGraw-Hill, New York, 1992.

8. James R. Bowerman, "An introduction to business element forecasting," *CMG '87 Conference Proceedings,* Computer Measurement Group, 1987, 703—709.

9. Paul Bratley, Bennett L. Fox, and Linus E. Schrage, *A Guide to Simulation,* Second Edition, Springer-Verlag, New York, 1987

10. Leroy Bronner, *Capacity Planning: Basic Hand Analysis*, IBM Washington Systems Center Technical Bulletin, December 1983.

11. Tim Browning, "Forecasting computer resources using business elements: a pilot study," *CMG '90 Conference Proceedings*, Computer Measurement Group, 1990, 421—427.

12. Jeffrey P. Buzen, *Queueing network models of multiprogramming*, Ph.D. dissertation, Division of Engineering and Applied Physics, Harvard University, Cambridge, MA, May 1971.

13. James D. Calaway, "SNAP/SHOT VS BEST/1." *Technical Support*, March 1991, 18—22.

14. C. Chatfield, *The Analysis of Time Series: An Introduction,* Third Edition, Chapman and Hall, London, 1984.

15. Edward I. Cohen, Gary M. King, and James T. Brady, "Storage hierarchies," *IBM Systems Journal*, 28(1), 1989, 62—76.

16. Peter J. Denning, "RISC architecture," *American Scientist*, January-February 1993, 7—10.

17. Hans Dithmar, Ian St. J. Hugo, and Alan J. Knight, *The Capacity Management Primer*, Computer Capacity Management Service Ltd., 1989. (Also available from the Institute for Computer Capacity Management.)

18. Jack Dongarra, Joanne L. Martin, and Jack Worlton, "Computer benchmarking: paths and pitfalls," *IEEE Spectrum,* July 1987, 38—43.

19. Brian Duncombe, "Managing your way to effective service level agreements," *Capacity Management Review*, December 1992 1–4.

20. Domenico Ferrari, Giuseppe Serazzi, and Alessandro Zeigner, *Measurement and Tuning of Computer Systems,* Prentice-Hall, Englewood Cliffs, NJ, 1983.

21. Randolph W. Hall, *Queueing Methods*, Prentice-Hall, Englewood Cliffs, NJ, 1991

22. Richard W. Hamming, *The Art of Probability for Scientists and Engineers*, Addison-Wesley, Reading, MA, 1991.

23. John L. Hennessy and David A. Patterson, *Computer Architecture: A Quantitative Approach,* Morgan Kaufmann, San Mateo, CA, 1990.

24. Phillip C. Howard, Editor, *IS Capacity Management Handbook Series, Volume 1, Capacity Planning,* Institute for Computer Capacity Management, updated every few months.

25. Phillip C. Howard, Editor, *IS Capacity Management Handbook Series, Volume 2, Performance Analysis and Tuning,* Institute for Computer Capacity Management, updated every few months.

26. IBM, *MVS/ESA Resource Measurement Facility Version 4 General Information,* GC28-1028-3, IBM, March 1991.

27. Thomas F. Incorvia, "Benchmark cost, risks, and alternatives," *CMG '92 Conference Proceedings*, Computer Measurement Group, 1992, 895—905.

28. William H. Inmon, D*eveloping Client/Server Applications, Revised Edition*, QED Publishing Group, Wellesley, MA, 1993.

29. David K. Kahaner and Ulrich Wattenberg, "Japan: a competitive assessment," *IEEE Spectrum*, September 1992, 42—47.

30. Peter J. B. King, *Computer and Communication Systems Performance Modelling*, Prentice-Hall, Hertfordshire, UK, 1990.

31. Leonard Kleinrock, *Queueing Systems, Volume I: Theory*, John Wiley, New York, 1975.

32. Leonard Kleinrock, *Queueing Systems, Volume II: Computer Applications*, John Wiley, New York, 1976.

33. Donald E. Knuth, T*he Art of Computer Programming: Seminumerical Algorithms,* Second Edition, Addison-Wesley, Reading, MA, 1981.

34. C. B. Kube, *TPNS: A Systems Test Tool to Improve Service Levels,* IBM Washington Systems Center, GG22-9243-00, 1981.

35. Shui F. Lam and K. Hung Chan, *Computer Capacity Planning: Theory and Practice,* Academic Press, San Diego, 1987.

36. Stephen S. Lavenberg, Ed., *Computer Performance Modeling Handbook,* Academic Press, New York, 1983.

37. Edward D. Lazowska, John Zahorjan, G. Scott Graham, and Kenneth C. Sevcik, *Quantitative System Performance: Computer system Analysis Using Queueing Network Models*, Prentice-Hall, Englewood Cliffs, NJ, 1984.

38. John D. C. Little, "A proof of the queueing formula: $L = \lambda W$," *Operations Res.*, 9(3), 1961, 383—387.

39. T. L. Lo and J. P. Elias, "Workload forecasting using NFU: a capacity planner's perspective," *CMG '86 Conference Proceedings,* Computer Measurement Group, 1986, 115—120.

40. M. H. MacDougall, *Simulating Computer Systems: Techniques and Tools*, The MIT Press, Cambridge, MA, 1987.

41. Edward A, MacNair and Charles H. Sauer, *Elements of Practical Performance Modeling,* Prentice-Hall, Englewood Cliffs, NJ, 1985.

42. E. Wainright Martin, Daniel W. DeHayes, Jeffrey A. Hoffer, and William C. Perkins, *Managing Information Technology: What Managers Need to Know*, Macmillan, New York, 1991.

43. H. W. "Barry" Merrill, *Merrill's Expanded Guide to Computer Performance Evaluation Using the SAS System*, SAS, Cary, NC, 1984.

44. George W. (Bill) Miller, "Service Level Agreements: Good fences make good neighbors," *CMG'87 Conference Proceedings*, Computer Measurement Group, 1987, 553—560.

45. George W. (Bill) Miller, "Workload characterization and forecasting for a large commercial environment," *CMG '87 Conference Proceedings,* Computer Measurement Group, 1987, 655—665.

46. Mark A. Miller, *Internetworking: A Guide to Network Communications LAN to LAN; LAN to WAN,* M&T Books, Redwood City, CA 1991.

47. Michael K. Molloy, *Fundamentals of Performance Modeling,* Macmillan, New York, 1989.

48. Stephen Morse, "Benchmarking the benchmarks," *Network Computing*, February 1993, 78—84.

49. Robert Orfali and Dan Harkey, *Client-Server Programming with OS/2 Extended Edition,* Second Edition, Van Nostrand Reinhold, New York, 1992,

50. David A. Patterson, Garth Gibson, Randy H. Katz, "A case for redundant arrays of inexpensive disks (RAID)," *ACM SIGMOD Conference Proceedings*, June 1—3, 1988, 109—116. Reprinted in *CMG Transactions, Fall 1991*.

51. Randolph W. Hall, *Queueing Methods for Services and Manufacturing,* Prentice-Hall, Englewood Cliffs, NJ, 1991.

52. Martin Reiser, “Mean value analysis of queueing networks, A new look at an old problem,” *Proc. 4th Int. Symp. on Modeling and Performance Evaluation of Computer Systems*, Vienna (1979).

53. Martin Reiser, “Mean value analysis and convolution method for queue-dependent servers in closed queueing networks,” *Performance Evaluation*, 1(1), January 1981, 7—18.

54. Martin Reiser and Stephen S. Lavenberg, “Mean value analysis of closed multichain queueing networks,” *JACM*, 22, April 1980, 313—322.

55. John M. Reyland, “The use of natural forecasting units,” *CMG ‘87 Conference Proceedings,* Computer Measurement Group, 1987, 710—13.

56. Thomas G. Robertazzi, *Computer Networks and Systems: Queueing Theory and Performance Evaluation*, Springer-Verlag, New York, 1990.

57. Jerry L. Rosenberg, “The capacity planning manager’s phrase book and survival guide,” *CMG ‘92 Conference Proceedings*, Computer Measurement Group, 1992, 983—989.

58. Omri Serlin, “MIPS, Dhrystones and other tales,” *Datamation,* June 1986, 112—118.

59. Carol Swink, “SPE in a client/server environment: a case study,” *CMG ‘92 Conference Proceedings*, Computer Measurement Group, 1992, 271—276.

60. Andrew S. Tanenbaum, *Computer Networks,* Second Edition, *Prentice-Hall,* Englewood Cliffs, NJ, 1988.

61. Michael Turner, Douglas Neuse, and Richard Goldgar, “Simulating optimizes move to client/server applications, *CMG ‘92 Conference Proceedings*, Computer Measurement Group, 1992, 805—812.

62. Reinhold P. Weicker, “An overview of common benchmarks,” *IEEE Computer*, December 1990, 65—75.

63. Peter D. Welch, “The statistical analysis of simulation results,” in *Computer Performance Modeling Handbook*, Stephen S. Lavenberg, Ed., Academic Press, New York, 1983.

64. Raymond J. Wicks, *Balanced Systems and Capacity Planning,* IBM Washington Systems Center Technical Bulletin GG22-9299-03, September 1991.

65. Kaisson Yen, “Projecting SPU capacity requirements: a simple approach,” *CMG ‘85 Conference Proceedings, Computer* Measurement Group, 1985, 386—391.

Appendix A Mathematica Programs

A.1 Introduction

Before we discuss the programs in the packages **work.m** and **first.m** we would like to warn you of some of the booby traps that exist in *Mathematica*, especially in Version 2.0 or 2.1. The trap that catches most users is called "conflicting names" by Nancy Blachman in her very useful book [Blachman 1992]. She discusses conflicting names in Section 15.6 on pages 256 through 258 of her book. Suppose you want to use the program **perform** from the package **first.m** but forget to load **first.m** before you try to use **perform**. As we show here, when you attempt to use **perform**, *Mathematica* merely echoes your request. After you load **first.m** and thus **perform**, the **perform** program works correctly. If you had attempted to use the program **Regress** from **LinearRegression** you would find an even more frustrating situation. You actually get a warning message on line 9 when you load the **LinearRegression** packages telling you that there is a conflict between the two versions of *Regress*. For some reason the warning message was not captured by the utility **SessionLog.m.** The exact warning message is:

Regress: Warning: Symbol Regress appears in multiple contexts {Statistics'LinearRegression', Global'}; definitions in context
Statistics'LinearRegression'
may shadow or be shadowed by other definitions.

The **Remove** command allows you to erase the global version of **Regress** so you can access the **LinearRegression** version of **Regress** as we show in the following *Mathematica* session segment, which is slightly scrambled because some of the output is too wide for the page.

```
In[4]:= perform[23, 45]

Out[4]= perform[23, 45]
```

```
In[5]:= <<first.m
In[6]:= perform[23,45]

Machine A is n% faster than machine B where n = 95.65217391

In[7]:= data = {{1, 2}, {2, 3.5}, {3, 4.2}}

Out[7]= {{1, 2}, {2, 3.5}, {3, 4.2}}

In[8]:= Regress[data, {1, x}, x]

Out[8]= Regress[{{1, 2}, {2, 3.5}, {3, 4.2}}, {1, x}, x]

In[9]:= <<Statistics`LinearRegression`

In[10]:= Regress[data, {1, x}, x]

Out[10]= Regress[{{1, 2}, {2, 3.5}, {3, 4.2}}, {1, x}, x]

In[11]:= Remove[Regress]

In[12]:= Regress[data, {1,x}, x]

Out[12]= {ParameterTable ->    Estimate   SE         TStat     PValue ,
                          1   1.03333    0.498888   2.07127   0.286344

                          x   1.1        0.23094    4.76314   0.131742

>  RSquared -> 0.957784, AdjustedRSquared -> 0.915567,

>  EstimatedVariance -> 0.106667,

>  ANOVATable ->        DoF   SoS        MeanSS     FRatio     PValue  }
             Model  1   2.42       2.42       22.6875    0.131742

             Error  1   0.106667   0.106667
             Total  2   2.52667
```

Sometimes, when you have loaded a number of packages the contexts can get so scrambled that you must sign off from *Mathematica* with the **Quit** command and start over again.

Version 2.0 of *Mathematica* provides a number of help messages that were not present in Version 1.2. These messages are sometimes very useful and at other times seem like useless nagging. The help messaging system gets very exercised if you use names that are similar. For example, if you type "function = 12", you will get the following message:

General::spell1:

Possible spelling error: new symbol name "function"

is similar to existing symbol "Function".

This may your first warning that **Function** is the name of a *Mathematica* function. You can get a similar message by entering "frank = 12" and "mfrank = 1". The warning message is:

General::spell1:

Possible spelling error: new symbol name "mfrank"

is similar to existing symbol "frank".

Messages like this can be a little annoying but come with the territory.

Abell and Braselton wrote two books about *Mathematica* which were published in 1992. In the first book, *Mathematica by Example*, they provide several examples of the use of the package **LinearRegression.m** as well as a number of other packages that come with Mathematica Version 2.0. In their second book, *Mathematica Handbook*, they provide even more discussion of the packages. Both of their books are heavily oriented toward the Macintosh *Mathematica* front end but provide a great many examples that can be appreciated by anyone with any version of *Mathematica*. At the time of this writing (June 1993) the Macintosh and Next Mathematica front ends are more elaborate than those for the various UNIX versions or the two versions for the IBM PC and compatibles. Rumors abound that the long-awaited X-Windows front end will be available soon.

The package stored in the file `first.m` and that stored in `work.m` follow.

```
BeginPackage["first`"]
first::usage="This is a collection of functions used in this book."
perform::usage="perform[A_, B_] calculates the percentage faster one machine
is over the other where A is the execution time on machine A and B is the execu-
```

```
tion time on machine B." speedup::usage="speedup[enhanced_, speedup_] calcu-
lates the speedup of an improvement where enhanced is the percentage of time in
enhanced mode while speedup is the speedup while in enhanced mode."
bounds::usage="bounds[numN_, think_, demands_] calculates the bound on
throughput and response time for a closed, single workload class queueing model
like that shown in Figure 3.4. Here numN is the number of terminals, think is the
average think time, and demands is a vector of the service demands."
nancy::usage="nancy[n_] provides my solution to Exercise 1.2."
trial::usage="trial[n_] is the program demonstrated in Example 1.1 to show that
Marilyn vos Savant gave the correct solution to the Monty Hall problem."
makeFamily::usage="makeFamily[ ] returns a list of children. This is one of
Nancy Blachman's programs used with her permission."
numChildren::usage="numChildren[n] returns statistics on the number of chil-
dren from n families. This is another of Nancy Blachman's programs used with
permission."  cpu::usage="cpu[instructions_, MHz_, cputime_] calculates the
speed in MIPS and the CPI for a cpu where  instructions is number of instructions
executed by the cpu in the length of time cputime where the CPU runs at the
speed (in MHz) of MHz."
simpledisk::usage="simpledisk[seek_, rpm_, dsectors_, tsectors_, controller_]
where seek is the  seek time in milliseconds, rpm is the revolutions per minute,
dsectors is the number of sectors per  track, tsectors is the number of sectors to be
transferred, and  controller  is the estimated controller time calculates the latency,
the transfer time, and the access time."


Begin["first`private`"]

perform[A_, B_] :=
(* A is the execution time on machine A *)
 (* B is the execution time on machine B *)
 Block[{n, m},
 n = ((B-A)/A) 100;
 m = ((A-B)/B) 100;
If[A <= B, Print["Machine A is n% faster than machine B where n = ", N[n, 10]],
Print["Machine B is n% faster than machine A where n = ", N[m, 10]]]; ]

speedup[enhanced_, speedup_] :=
(* enhanced is percent of time in enhanced mode *)
(* speedup is speedup while in enhanced mode *)
Block[{frac, speed},
frac = enhanced / 100;
speed = 1 /(1 - frac + frac / speedup);
Print["The speedup is ", N[speed, 8]]; ]
```

```
bounds[numN_, think_, demands_]:=
Block[{m, dmax,d, j},
m=Length[demands];
dmax=Max[demands];
d=Sum[demands[[j]], {j, 1, m}];
lowerx=numN/(numN d+think);
upperx=Min[numN/(d+think),1/dmax];
lowerr=Max[d, numN dmax-think];
upperr=numN d;
Print["Lower bound on throughput is  ",lowerx];
Print["Upper bound on throughput is  ",upperx];
Print["Lower bound on response time is ",lowerr];
Print["Upper bound on response time is ",upperr]; ]

nancy[n_] :=
Block[{i,trials, average,k},
(* trials counts the number of births *)
(* for each couple. It is initialized to zero. *)
trials=Table[0, {n}];
For[i=1, i<=n, i++,
While[True,trials[[i]]=trials[[i]]+1;
If[Random[Integer, {0,1}]>0,  Break[]] ];];
(* The while statement counts the number of births for couple i. *)
(* The while is set up to test after a pass through the loop *)
(* so we can count the birth of the first girl baby. *)
average=Sum[trials[[k]], {k, 1, n}]/n;
Print["The average number of children is  ", average];
 ]

trial[n_] :=
Block[{switch=0, noswitch=0},
correctdoor=Table[Random[Integer, {1,3}], {n}];
firstchoice=Table[Random[Integer, {1,3}], {n}];
For[i=1, i<=n, i++,
If[Abs[correctdoor[[i]]-firstchoice[[i]]]>0,
switch=switch+1, noswitch=noswitch+1]];
Return[{N[switch/n,8],N[noswitch/n,8]}];
]

makeFamily[]:=
   Block[{
         children = { }
```

```
      },
      While[Random[Integer] == 0,
         AppendTo[children, "girl"]
   ];
      Append[children, "boy"]
 ]
makeFamily::usage="makeFamily[ ] returns a list of children."
numChildren[n_Integer] :=
  Block[{
        allChildren
       },
       allChildren = Flatten[Table[makeFamily[ ], {n}]];
       {
        avgChildren -> Length[allChildren]/n,
        avgBoys -> Count[allChildren, "boy"]/n,
        avgGirls -> Count[allChildren, "girl"]/n
       }
   ]
numChildren::usage="numchildren[n] returns statistics on
     the number of children from n families."

cpu[instructions_, MHz_, cputime_] :=
(* instructions is number of instructions executed by *)
(* the cpu in the length of time cputime *)
Block[{cpi,mips},
mips = 10^(-6) instructions / cputime;
cpi = MHz / mips;
Print["The speed in MIPS is ", N[mips, 8]];
Print["The number of clock cycles per instruction, CPI, is ", N[cpi,10]]; ]

simpledisk[seek_, rpm_, dsectors_, tsectors_, controller_] :=
Block[{latency, transfer},
(* seek time in milliseconds, dsectors is number of sectors per *)
(* track, tsectors is number of sectors to be transferred *) (* controller is esti-
mated controller time *)
Block[{latency, transfer, access},
latency = 30000/rpm;
transfer = 2 latency tsectors / dsectors;
access = latency + transfer + seek + controller;
Print["The latency time in milliseconds is ", N[latency, 5]];
Print["The transfer time in milliseconds is ", N[transfer, 6]];
Print["The access time in milliseconds is ", N[access, 6]]; ]] End[]
```

```
(* end 'private' context *)
EndPackage[]
```

```
BeginPackage["work'", "Statistics'NormalDistribution'", "Statistics'Com-
mon'DistributionsCommon'", "Statistics'ContinuousDistributions'"]
work::usage="This is a collection of functions used in this book."
sopen::usage="sopen[lambda_, v_?VectorQ, s_?VectorQ ] computes the perfor-
mance statistics for the single workload class open model of Figure 4.1. For this
program lambda is the average throughput, v is the vector of visit ratios for the
service centers, and s is the vector of the average service time per visit for each
service center."
sclosed::usage="sclosed[N_?IntegerQ, D_?VectorQ, Z_] computes the perfor-
mance statistics for the single workload class closed model of Figure 4.2. N is the
number of terminals, D is the vector of service demands and Z is the mean think
time at each terminal."
mopen::usage="mopen[lambda_, d_ ] computes the performance statistics for the
multiple workload class open model of Figure 4.1. For this program lambda is the
average throughput and d is the C by K matrix of service demands."
cent::usage="cent[N_?IntegerQ, D_?VectorQ] computes the performance statis-
tics for the central server
model with fixed MPL N. D is the service demand vector."
online::usage="online[m_?IntegerQ, Demands_?VectorQ, N_?IntegerQ, T_]
computes the performance statistics for a terminal system with a FESC to replace
the central server model of the computer system. The program subcent is used to
calculate the rates needed as input. The maximum multiprograming level allowed
is m. Demands is the vector of service demands. N is the number of active termi-
nals or workstations connected to the computer system. T is the mean think time
for the users at the terminals."
subcent::usage="Computes the throughput for a central server model with fixed
MPL."
Exact::usage="Exact[Pop_?VectorQ, Think_?VectorQ, Demands_?MatrixQ ]
computes the performance statistics for the closed multiworkload class model of
Figure 4.2. Pop is the vector of population by class. Think is the vector of think
time by class and Demands is the C by K matrix of service demands."
Approx::usage="Approx[Pop_?VectorQ, Think_?VectorQ, Demands_?MatrixQ,
epsilon_Real ] computes the performance statistics for the closed multiworkload
class of Figure 4.2 using an approximation technique. Pop is the vector of popu-
lation by class. Think is the vector of think time by class and Demands is the C
by K matrix of service demands. The parameter epsilon determines how accu-
rately the algorithm attempts to calculate the solution."
```

mm1::usage="mm1[lambda_, es_] calculates the performance statistics for the classical N/M/1 queueing model where lambda is the average arrival rate and es is the average service time."
simmm1::usage="simmm1[lambda_Real, serv_Real, seed_Integer, n_Integer, m_Integer] uses simulation to compute the average time in the system and as well as the 95th percent confidence interval for it using the method of batched means. The parameters lambda and serv are the average arrival rate and the average service time of the server. The parameter seed determines the sequence of random numbers used in the model, n is the number of customers used in the warmup run and m is the number of customers served in each of the 100 subruns."
chisquare::usage="chisquare[alpha_, x_, mean_] uses Knuth's algorithm to test the hypothesis that the vector of numbers x is a sample from an exponential distribution with average value mean at the alpha level of significance."
ran::usage="ran[a_Integer, m_Integer, n_Integer, seed_Integer] computes n random integers using the linear congruential method with parameters a and m beginning with the seed."
uran::usage="uran[a_Integer, m_Integer, n_Integer, seed_Integer] generates a uniform random sequence between zero and one."
rexpon::usage="rexpon[a_Integer, m_Integer, n_Integer, seed_Integer, mean_-Real] generates a random sequence of n exponentially distributed numbers with average value mean."
Fixed::usage="Fixed[Ac_, Nc_, Zc_, Dck_, epsilon_Real] is a program bases on Approx that will work with fixed throughput classes. It is described in Section 4.2.2.4."
pri::usage="Pri[Pop_?VectorQ, Think_?VectorQ, Demands_?MatrixQ, epsilon_Real] computes the performance statistics for the closed multiworkload class of Figure 4.2 with priorities."

```
Begin["work`private`"]
mopen[ lambda_, d_ ] :=
(* multiple class open queueing model *)
Block[ {u, R, r, L, u1, C, K, u2},
 u=lambda * d;
 C = Length[lambda];
 K = Length[d[[2]]];
 u1=Apply[Plus, u];
 x=1/(1-u1);
 r = Transpose[x Transpose[d]];
 l = lambda r;
 R= Apply[Plus, Transpose[r]];
 L=lambda R;
 number = Apply[Plus, l];
 Print[ "" ] ;
```

```
Print[ "" ] ;
Print[
SequenceForm[
ColumnForm[ Join[ {"Class#", "------"}, Range[C] ], Right ],
ColumnForm[ Join[ {" TPut", " ----------"}, SetAccuracy[ lambda, 6 ] ], Right ],
ColumnForm[ Join[ {" Number", " ---------"}, SetAccuracy[ L, 6 ] ], Right ],
ColumnForm[ Join[ {" Resp", " --------------"}, SetAccuracy[R,6]], Right ]]];
Print[ "" ] ;
Print[ "" ] ;
Print[
SequenceForm[
ColumnForm[ Join[ {"Center#", "------"}, Range[K] ], Right ],
ColumnForm[ Join[ {" Number", " ---------"}, SetAccuracy[ number, 6 ] ], Right
],
ColumnForm[ Join[ {" Utiliz", " ----------"}, SetAccuracy[u1,6]], Right ]]];
]

sopen[ lambda_, v_?VectorQ, s_?VectorQ ] :=
(* single class open queueing model *)
Block[ {n, d, dmax, xmax, u, u1, k},
d = v s ;
dmax=Max[d];
xmax=1/dmax;
u=lambda*d;
x=lambda*v;
numK = Length[v];
r=d/(1-u);
l=lambda*r;
R=Apply[Plus, r];
L=lambda*R;
Print[];
Print["The maximum throughput is ",N[xmax, 6]];
Print["The system throughput is ", N[lambda, 6]];
Print["The system mean response time is ",N[R, 6]];
Print["The mean number in the system is ",N[L, 6]];
Print[ "" ] ;
Print[ "" ] ;
Print[
SequenceForm[
ColumnForm[ Join[ {"Center#", "------"}, Range[numK] ], Right ],
ColumnForm[ Join[ {" Resp", " ----------"}, SetAccuracy[ r, 6 ] ], Right ],
ColumnForm[ Join[ {" TPut", " ---------"}, SetAccuracy[ x, 6 ] ], Right ],
```

```
 ColumnForm[ Join[ {" Number", " ---------"}, SetAccuracy[ l, 6 ] ], Right ],
 ColumnForm[ Join[ {" Utiliz", " -------"}, SetAccuracy[u, 6]], Right ]]];
 ]
```

```
sclosed[N_?IntegerQ, D_?VectorQ, Z_]:=
(* Single class exact closed model *)
Block[{L, r, n, X, u, l, R, K},
K = Length[D];
l=Table[0, {K}];
r=Table[0, {K}];
For[n=1, n<=N, n++, r=D*(l+1); R=Apply[Plus,r]; X=n/(R+Z);
l=X r; u=X D];
l = X r;
L=X R;
numK = K;
su = u;
 Print[""];
 Print[""];
 Print["The system mean response time is ", R];
 Print["The system mean throughput is ", X];
 Print["The average number in the system is ", L];
 Print[ "" ] ;
 Print[ "" ] ;
 Print[
 SequenceForm[
 ColumnForm[ Join[ {"Center#", "------"},Range[numK ] ], Right ],
 ColumnForm[ Join[ {" Resp", " ----------"}, SetAccuracy[ r, 6 ] ], Right ],
 ColumnForm[ Join[ {" Number", " ---------"}, SetAccuracy[ l, 6 ] ], Right ],
 ColumnForm[ Join[ {" Utiliz", " -----------"}, SetAccuracy[su, 6]], Right ]]];
 ]
```

```
cent[N_?IntegerQ, D_?VectorQ]:=
(* central server model *)
(* k is number of service centers *)
(* N is MPL, D is service demand vector *)
Block[{L, w,k, wn, n, lambdan, rho},
k = Length[D];
L = Table[0, {k}];
For[n=1, n<=N, n++, w=D*(L+1); wn=Apply[Plus,w]; lambdan=n/wn;
```

```
L=lambdan w; rho=lambdan D];
(* lambdan is mean throughput *)
(* wn is mean time in system *)
(* L is vector of number at servers *)
(* rho is vector of utilizations *)
 Print[""];
 Print[""];
 Print["The average response time is ", wn];
 Print["The average throughput is ", lambdan];
 Print[ "" ] ;
 Print[ "" ] ;
 Print[
 SequenceForm[
 ColumnForm[ Join[ {"Center#", "------"}, Range[k] ], Right ],
 ColumnForm[ Join[ {" Number", " ---------"}, SetAccuracy[L, 6 ] ], Right ],
 ColumnForm[ Join[ {" Utiliz", " ----------"}, SetAccuracy[rho,6]], Right ]]];
 ]

online/: online[m_?IntegerQ,Demands_?VectorQ , N_?IntegerQ, T_]:=
(* N is the number of active terminals or workstations connected *)
(* to the computer system. T is the mean think time for the *)
(* users at the terminals. The maximum multiprograming *)
(* level allowed is m. *)
Block[{n, w,s, L, r, x, nsrate, q, q0, },
r = srate[m, Demands];
r = Flatten[r];
x=Table[Last[r], {N-m}];
nsrate=Join[r, x];
q=Join[{1}, Table[0, {N-1}]];
s=0;
q0=1;
For[n=1, n<=N, n++,
w=0;
For[j=1, j<=n, j++,
w=w+(j/nsrate[[j]])*If[j>1, q[[j-1]], q0];
lambda=n/(T+w)];
s=0;
For[j=n, j>=1, j--,
q[[j]]=(lambda/nsrate[[j]])*If[j>1, q[[j-1]],q0];
s=s+q[[j]]];
q0=1-s
];
```

```
L = lambda w;
qplus=Join[{q0},q];
probin = Flatten[{Take[qplus, m], 1 - Apply[Plus, Take[qplus, m]]}];
numberin = Drop[probin, 1] . Range[1,m];
timein = numberin / lambda;
numberinqueue = L - numberin;
timeinqueue = numberinqueue / lambda;
U = lambda * Demands;
k = Length[Demands];
(* lambda is mean throughput *)
(* w is mean response time *)
(* qplus is vector of conditional probabilities *)
 Print[""];
 Print[""];
 Print["The average number of requests
 in process is ", L];
 Print["The average system throughput is ", lambda];
 Print["The average system response time is ", w];
 Print["The average number in main memory is ", SetAccuracy[numberin,5]];
 Print[ "" ] ;
 Print[ "" ] ;
 Print[
 SequenceForm[
 ColumnForm[ Join[ {"Center#", "------"}, Range[k] ], Right ],
 ColumnForm[ Join[ {" Utiliz", " ----------"}, SetAccuracy[U,6]], Right ]]];
 ]
```

```
subcent[k_?IntegerQ,N_?IntegerQ, D_?VectorQ]:=
(* central server model *)
(* k is number of service centers *)
(* N is MPL, D is service demand vector *)
Block[{L, w, wn, n, lambdan, rho},
L=Table[0, {k}];
For[n=1, n<=N, n++, w=D*(L+1); wn=Apply[Plus,w]; lambdan=n/wn;
L=lambdan w; rho=lambdan D];
(* lambdan is mean throughput *)
(* wn is mean time in system *)
(* L is vector of number at servers *)
(* rho is vector of utilizations *)
```

```
Return[{lambdan}];
]
 srate[m_?IntegerQ, Demands_?VectorQ] :=
 Block[{n },
 k = Length[Demands];
 rate = {};
 For[n = 1, n<=m, n++, rate = Join[ rate, subcent[k, n, Demands]]];
 Return[{rate}];
 ]


FixPerm[ numC_, v_, Pop_ ] :=
Block[ {x, m = v },
 For[x=numC, x>1, x--,
 If[m[[x]] > Pop[[x]],
 m[[x-1]]=m[[x-1]]+m[[x]]-Pop[[x]] ;
 m[[x]]=Pop[[x]] ]];
 If[ m[[1]] > Pop[[1]], {}, m]
]

FirstPerm[ numC_, Pop_, n_ ] :=
Block[ {m},
 m = Table[ 0, { numC } ] ;
 m[[numC]] = n ;
 FixPerm[numC, m, Pop ]
]

NextPerm[ numC_, Pop_, v_ ] :=
Block[ {m=v, x=numC, y},

 While[ m[[x]] == 0, x-- ] ;
 If[x==1, Return[{}] ] ;

 m[[x]]-- ;
 x-- ;
 While[ (x >= 1) && (m[[x]] == Pop[[x]]), x-- ] ;
 If[x < 1, Return[{}] ] ;
 m[[x]]++ ;

 For[y=x+1, y<numC, y++,
 m[[numC]] = m[[numC]] + m[[y]] ;
 m[[y]] = 0 ] ;
```

```
 FixPerm[numC, m, Pop ]
]



Exact[ Pop_?VectorQ, Think_?VectorQ, Demands_?MatrixQ ] :=

Block[ { n, c, k, v, nMinus1, r, cr, x, q1, q2, qtemp, su, sq,
 numC = Length[Pop],
 numK = Dimensions[Demands][[2]],
 totalP, zVectorK },

 zVectorK = Table[0, {numK}] ;
 totalP = Sum[ Pop[[i]], {i, 1, numC} ] ;
 q1[ Table[0, {numC}] ] = zVectorK ;

 For[n=1, n <= totalP, n++,
 v = FirstPerm[numC, Pop, n ] ;
 While[v!= {},
 r = Table[(nMinus1 = v ;
 If[ nMinus1[[i]] > 0,
 nMinus1[[i]]-- ;
 Demands[[i]] * ( 1 +
 If[OddQ[n], q1[nMinus1], q2[nMinus1]]), zVectorK]),
 { i, 1, numC} ] ;

 x = Think + Apply[Plus, r, {1} ] ;
 For[c=1, c<=numC, c++, If[ x[[c]]>0, x[[c]] = v[[c]] / x[[c]] ] ] ;

 qtemp = x . r ;
 If[OddQ[n], q2[v]=qtemp, q1[v]=qtemp ] ;

 v = NextPerm[numC, Pop, v ] ];
 If[OddQ[n], Clear[q1], Clear[q2]]
 ] ;

 cr = Apply[Plus, r, 1 ] ;
 su = x . Demands ;
 l = x . r ;

 Print[ "" ] ;
 Print[ "" ] ;
```

```
Print[
SequenceForm[
ColumnForm[ Join[ {"Class#", "------"},Range[numC] ], Right ],
ColumnForm[ Join[ {" Think", " -----"}, Think ], Right ],
ColumnForm[ Join[ {" Pop", " -----"}, Pop ], Right ],
ColumnForm[ Join[ {" Resp", " ---------"}, SetAccuracy[ cr, 6 ] ], Right ],
ColumnForm[ Join[ {" TPut", " ---------"}, SetAccuracy[ x, 6 ] ], Right ] ] ];


Print[ "" ] ;
Print[ "" ] ;
Print[
SequenceForm[
ColumnForm[ Join[ {"Center#", "------"},Range[numK ] ], Right ],
ColumnForm[ Join[ {" Number", " ---------"}, SetAccuracy[ l, 6 ] ], Right ],
ColumnForm[ Join[ {" Utiliz", " -----------"}, SetAccuracy[su, 6]], Right ]]];

]

Approx[ Pop_?VectorQ, Think_?VectorQ, Demands_?MatrixQ, epsilon_Real ]
:=

Block[ { Flag, a, r, x, newQueueLength, qTot, q, cr, sq, su, it,
 numC = Length[Pop],
 numK = Dimensions[Demands][[2]],t, number },

 q = N[Table[ Pop[[c]]/numK, {c, 1, numC}, {k, 1, numK} ] ] ;

 Flag = True ;
 While[Flag==True,

 qTot = Apply[Plus, Transpose[q], 1] ;
 a = Table[ qTot[[k]] - q[[c,k]] + ((Pop[[c]]-1)/Pop[[c]]) q[[c,k]],
 {c, 1, numC}, {k, 1, numK} ] ;

 r = Table[ Demands[[c,k]] (1 + a[[c,k]]), {c, 1, numC}, {k, 1, numK} ] ;
 cr = Apply[Plus, r, 1 ] ;
 x = Pop / ( Think + cr) ;

 Flag = False ;
 q = Table[(newQueueLength = x[[c]] r[[c,k]] ;
 If[ Abs[ q[[c,k]] - newQueueLength ] >= epsilon, Flag=True] ;
```

```
newQueueLength),
{c, 1, numC}, {k, 1, numK} ] ;
] ;

(* Compute final results *)

su = x . Demands ;
number = x . r ;


Print[ "" ] ;
Print[ "" ] ;
Print[
SequenceForm[
ColumnForm[ Join[ {"Class#", "------"}, Table[ c, {c,1,numC} ] ], Right ],
ColumnForm[ Join[ {" Think", " -----"}, Think ], Right ],
ColumnForm[ Join[ {" Pop", " -----"}, Pop ], Right ],
ColumnForm[ Join[ {" Resp", " -------------"}, SetAccuracy[ cr, 6 ] ], Right ],
ColumnForm[ Join[ {" TPut", " -----------"}, SetAccuracy[ x, 6 ] ], Right ] ] ] ;

Print[ "" ] ;
Print[ "" ] ;
Print[
SequenceForm[
ColumnForm[ Join[ {"Center#", "------"}, Table[ c, {c,1,numK} ] ], Right ],
ColumnForm[ Join[ {"number", "--------------"}, SetAccuracy[number, 6]],
Right ],
ColumnForm[ Join[ {" Utilization", " -----------"}, SetAccuracy[su, 6]], Right
]]];
] /; Length[Pop] == Length[Think] == Length[Demands]


Fixed[ Ac_, Nc_,
Zc_, Dck_, epsilon_Real ] :=

Block[ { Flag, Rck, Xc, newQ, Qck, Rc, Qk, Uk, Pc, Tc,
numC = Length[Nc], numK = Dimensions[Dck][[2]] },

Tc = N[ Zc + Apply[Plus, Dck, 1 ] ] ;
Pc = N[ Table[ If[NumberQ[ Nc[[c]] ], Nc[[c]],
If[Zc[[c]]==0, 1, 100] ], {c, 1, numC} ] ];
Qck = Table[ Dck[[c,k]] / Tc[[c]] Pc[[c]], {c, 1, numC}, {k, 1, numK} ] ;
```

```
Flag = True ;
While[Flag==True,

Qk = Apply[Plus, Qck ] ;
Rck = Table[ Dck[[c,k]] *
(1 + Qk[[k]] - Qck[[c,k]] + Qck[[c,k]] *
If[ Pc[[c]] < 1, 0, ((Pc[[c]]-1)/Pc[[c]])] ),
{c, 1, numC}, {k, 1, numK} ] ;
Rc = Apply[Plus, Rck, 1 ] ;

Xc = Table[If[NumberQ[Ac[[j]]], Ac[[j]], Pc[[j]] / (Zc[[j]] + Rc[[j]])],
{j, 1, numC} ] ;
Pc = Table[If[NumberQ[Ac[[c]]], Xc[[c]] * (Zc[[c]] + Rc[[c]]), Pc[[c]] ],

{c, 1, numC} ] ;

Flag = False ;
Qck = Table[(newQ = Xc[[c]] Rck[[c,k]] ;
If[ Abs[ Qck[[c,k]] - newQ ] >= epsilon, Flag=True] ;
newQ), {c, 1, numC}, {k, 1, numK} ] ;
] ;

(* Compute final results *)

Uk = Xc . Dck ;
Qk = Xc . Rck ;


Print[ "" ] ;
Print[ "" ] ;
Print[
SequenceForm[
ColumnForm[ Join[ {"Class#", "--------"}, Table[ c, {c,1,numC} ] ], Right ],
ColumnForm[ Join[ {" ArrivR", " --------------"}, Ac ], Right ],
ColumnForm[ Join[ {" Pc", " ------------"}, Pc ], Right ]]];

Print[ "" ] ;
Print[ "" ] ;
Print[
SequenceForm[
ColumnForm[ Join[ {"Class#", "--------"}, Table[ c, {c,1,numC} ] ], Right ],
```

```
 ColumnForm[ Join[ {" Resp", " --------------"}, SetAccuracy[ Rc, 6 ] ], Right ],
 ColumnForm[ Join[ {" TPut", " -------------"}, SetAccuracy[ Xc, 6 ] ], Right ] ] ]
;
 Print[ "" ] ;
 Print[ "" ] ;
 Print[
 SequenceForm[
 ColumnForm[ Join[ {"Center#", "---------"}, Table[ c, {c,1,numK} ] ], Right ],
 ColumnForm[ Join[ {"Number", " -------------"}, SetAccuracy[Qk, 6]], Right ],
 ColumnForm[ Join[ {" Utiliz", " ---------"}, SetAccuracy[Uk, 6]], Right ]]];
]

Pri[ Pop_?VectorQ, Think_?VectorQ, Demands_?MatrixQ, epsilon_Real ] :=

Block[ { Flag, a, r, x, newQueueLength, qTot, q, cr, sq, su, it,
 numC = Length[Pop],
 numK = Dimensions[Demands][[2]] },

 q = N[Table[ Pop[[c]]/numK, {c, 1, numC}, {k, 1, numK} ] ] ;
 r = q ;

 Flag = True ;
 While[Flag==True,

 cr = Apply[Plus, r, 1 ] ;
 x = Pop / ( Think + cr) ;

 a = Table[ ((Pop[[c]]-1)/Pop[[c]]) q[[c,k]],
 {c, 1, numC}, {k, 1, numK} ] ;

 u = Table[ Demands[[c,k]] x[[c]], {c, 1, numC}, {k, 1, numK} ] ;

 DI = Table[ 1 - Sum[ u[[j,k]], { j, 1, c-1 } ], {c, 1, numC}, {k, 1, numK} ] ;

 r = Table[ Demands[[c,k]] (1 + a[[c,k]]) / DI[[c,k]], {c, 1, numC}, {k, 1, numK}
] ;


 cr = Apply[Plus, r, 1 ] ;
 x = Pop / ( Think + cr) ;

 Flag = False ;
```

```
 q = Table[(newQueueLength = x[[c]] r[[c,k]] ;
 If[ Abs[ q[[c,k]] - newQueueLength ] >= epsilon, Flag=True] ;
 newQueueLength),
 {c, 1, numC}, {k, 1, numK} ] ;
 ] ;

(* Compute final results *)

 cr = Apply[Plus, r, 1 ] ;
 x = Pop / ( Think + cr) ;

 utilize = x . Demands ;
 number = x . r ;

Print[ "" ] ;
 Print[ "" ] ;
 Print[
 SequenceForm[
 ColumnForm[ Join[ {"Class#", "------"}, Range[numC] ], Right ],
 ColumnForm[ Join[ {" Think", " -----"}, Think ], Right ],
 ColumnForm[ Join[ {" Pop", " ---------"}, Pop ], Right ],
 ColumnForm[ Join[ {" Resp", " -------------"}, SetAccuracy[ cr, 6 ] ], Right ],
 ColumnForm[ Join[ {" TPut", " ------------"}, SetAccuracy[ x, 6 ] ], Right ] ] ] ;
Print[ "" ] ;
 Print[ "" ] ;
 Print[
 SequenceForm[
 ColumnForm[ Join[ {"Center#", "---------"}, Table[ c, {c,1,numK} ] ], Right ],
 ColumnForm[ Join[ {"Number", " -------------"}, SetAccuracy[number, 6]],
Right ],
 ColumnForm[ Join[ {" Utiliz", " ---------"}, SetAccuracy[utilize, 6]], Right ]]];


] /; Length[Pop] == Length[Think] == Length[Demands]


mm1[lambda_, es_] :=
Block[{wq, rho, w, l, lq, piq90, piw90},
rho=lambda es;
w =es/(1-rho);
wq =rho w;
l=lambda w;
```

```
lq=lambda wq;
piq90=N[Max[w Log[10 rho], 0], 10];
piw90=N[w Log[10], 10];
Print[];
Print["The server utilization is ", rho];
Print["The average time in the queue is ", wq];
Print["The average time in the system is ",w];
Print["The average number in the queue is ",lq];
Print["The average number in the system is ",l];
Print["The average number in a nonempty queue is ",1/(1-rho)];
Print["The 90th percentile value of q is ",piq90];
Print["The 90th percentile value of w is ",piw90]
]

simmm1[lambda_Real, serv_Real, seed_Integer, n_Integer, m_Integer]:=
Block[{t1, t2, s, s2, t, i, j, k, lower, upper, v, w, h},
SeedRandom[seed];
t1=0;
t2=0;
s2=0;
For[w=0; i = 1, i<=n, i++,
s = - serv Log[Random[]];
t = - (1/ lambda) Log[Random[]];
If[w<t, w = s, w = w + s -t];
s2 = s2 + w];
Print["The mean value of time in system at end of warmup is ", N[s2/n, 5]];
t1=0;
t2=0;
For[j=1, j<=100, j++,
s2=0;
For[k=1, k<=m, k++,
t = - (1/lambda) Log[Random[]];
s = - serv Log[Random[]];
If[w<t, w =s, w = w + s - t];
s2 = s2 + w];
t1 = t1 + s2/m;
t2 = t2 + (s2/m)^2];
v = (t2 - (t1^2)/100)/99;
h = 1.984217 Sqrt[v]/10;
lower = t1/100 - h;
upper = t1/100 + h;
Print["Mean time in system is ",N[t1/100, 6]];
```

```
Print["95 percent confidence interval is"];
Print[lower, " to " ,upper];
]

chisquare[alpha_, x_, mean_]:=
Block[{n, y, xbar, x25, x50, x75, o, e, m, first},
chisdist = ChiSquareDistribution[3];
n= Length[x];
y = Sort[x];
(* We calculate the quartile values assuming x is exponential. *)
x25 = - mean Log[0.75];
x50 = -mean Log[0.5];
x75 = -mean Log[0.25];
o = Table[0, {4}];
o[[1]] = Length[Select[y, # <= x25 &]];
o[[2]] = Length[Select[y, x25 < # && # <= x50 &]];
o[[3]] = Length[Select[y, x50 < # && # <= x75 &]];
o[[4]] = Length[Select[y, # > x75 &]];
(* o is the observed number in each quarter defined by *)
(* the quartiles. *)
m = n/4;
e = Table[m, {4}];
(* e is the expected number in each quarter. One-fourth *)
(* in each. *)
first = ((o - e)^2)/m;
chisq = N[Apply[Plus,first], 6];
(* This is the chisq value. *)
q = CDF[chisdist, chisq];
(* q is the probability that any observed chisq value *)
(* will not exceed the value just observed *)
(* if x is exponential. *)
p = 1 - q;
(* p is the probability any value of chisq will be *)
(* greater than or equal to that just observed *)
(* if x is exponential. *)
Print["p is ", N[p, 6]];
Print["q is ", N[q, 6]];
If[p < alpha/2, Return[Print["The sequence fails because chisq is too large."]]];
If[q < alpha/2, Return[Print["The sequence fails because chisq is too small."]]];
If[p >= alpha/2 && q >= alpha/2, Return[Print["The sequence passes the test."]]]
]
```

```
ran[a_Integer, m_Integer, n_Integer, seed_Integer]:=
Block[{i},
output =Table[0, {n}];
output[[1]]=Mod[seed, m];
For[i = 2, i<=n, i++,
output[[i]] = Mod[a output[[i-1]], m]];
Return[output];
]

uran[a_Integer, m_Integer, n_Integer, seed_Integer]:=
Block[{i},
random = ran[a, m, n, seed];
output =Table[0, {n}];
output[[1]]=Mod[seed, m]/m;
For[i = 2, i<=n, i++,
output[[i]] = random[[i]]/m];
Return[output];
]

rexpon[a_Iteger, m_Integer, n_Integer, seed_Integer, mean_Real]:=
Block[{i,random, output},
random = uran[a, m, n, seed];
output =Table[0, {n}];
For[i =1, i<=n, i++,
output[[i]] = - mean Log[random[[i]]]];
Return[N[output, 6]];
]
End[]
EndPackage[]
```

A.2 References

1. Martha L. Abell and James P. Braselton, *Mathematica by Example*, Academic Press, Boston, 1992.

2. Martha L. Abell and James P. Braselton, *Mathematica Handbook*, Academic Press, Boston, 1992.

3. Nancy Blachman, *Mathematica: A Practical Approach*, Prentice Hall, Englewood Cliffs, NJ, 1992.

Index

D

E

F

G

H

I

J

K

L

M

N

O

P

Q

R

S

W

Y

Z